Variation-Aware Design of Custom Integrated Circuits: A Hands-on Field Guide

T0135383

Trent McConaghy · Kristopher Breen
Jeffrey Dyck · Amit Gupta

Variation-Aware Design of Custom Integrated Circuits: A Hands-on Field Guide

With Foreword by James P. Hogan

Trent McConaghy
Solido Design Automation Inc.
Saskatoon, SK
Canada

Jeffrey Dyck
Solido Design Automation Inc.
Saskatoon, SK
Canada

Kristopher Breen
Solido Design Automation Inc.
Saskatoon, SK
Canada

Amit Gupta
Solido Design Automation Inc.
Saskatoon, SK
Canada

ISBN 978-1-4899-9673-2   ISBN 978-1-4614-2269-3 (eBook)
DOI 10.1007/978-1-4614-2269-3
Springer New York Heidelberg Dordrecht London

Printed on acid-free paper

Springer is part of Springer Science+Business Media (www.springer.com)

Book organization, in the context of a variation-aware design flow:

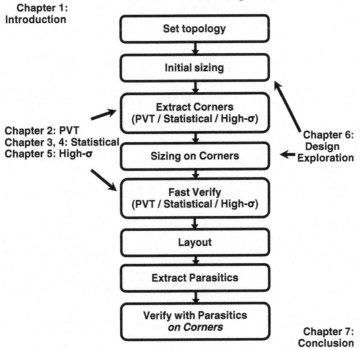

Chapter 1:
Introduction

Set topology

Initial sizing

Extract Corners
(PVT / Statistical / High-σ)

Chapter 2: PVT
Chapter 3, 4: Statistical
Chapter 5: High-σ

Sizing on Corners

Chapter 6:
Design
Exploration

Fast Verify
(PVT / Statistical / High-σ)

Layout

Extract Parasitics

Verify with Parasitics
on Corners

Chapter 7:
Conclusion

Foreword

After more than 35 years in the semiconductor business, I find that custom integrated circuit (IC) design continues to present extremely interesting challenges. In this book, Trent McConaghy, Kristopher Breen, Jeff Dyck, and Amit Gupta, all with Solido Design Automation, address the increasingly difficult design issues associated with variation in advanced nanoscale processes.

The authors have put together what I believe will become an invaluable reference for best practices in variation-aware custom IC design. They have taken theory and combined it with methodology and examples, based on their experiences in supplying leading-edge design tools to the likes of TSMC and NVIDIA. This book's content is useful for circuit designers, CAD managers and CAD researchers. This book will also be very useful to graduate students as they begin their careers in custom IC design.

I have always felt that the job of a designer is to optimize designs to what the requirements dictate, within the process capabilities. Circuits must trade off the marketing requirements of function, performance, cost, and power. This is especially true for custom IC design, where the results are on a continuum. There is never a perfect answer—only the most right, or equivalently, the least wrong.

Moore's Law—the practice of shrinking transistor sizes over time—has traditionally been a no-brainer, since smaller devices directly led to improved power, performance, and area. Several decades into Moore's Law, today's IC manufacturing has literally reached the level of "nanotech", with minimum device sizes at 40, 28, 20 nm, and most recently 14 nm. Variation in devices during manufacturing has always been around, but it has not traditionally been a big issue. The problem is that variation gets exponentially worse as the devices shrink, and it has become a major problem at these nano nodes. Designers must choose between over-margining so that the circuit yields (taking a performance hit), or to push performance (taking a yield hit).

Variation has made it harder to differentiate ICs on power or performance, while still yielding. The use of common commercial foundries, such as TSMC, GLOBALFOUNDRIES, Samsung, and even now Intel, makes it even more difficult to differentiate competitively. Performance hits are unacceptable, because all semiconductor companies are using the same foundries trying to produce competitive chips. In turn, yield hits are unacceptable for high volume applications since costs quickly skyrocket.

Moore's Law and opportunities for differentiation are the lifeblood of a healthy semiconductor industry. Variation is threatening both.

I've known Trent for several years now, since when he was co-founder of Analog Design Automation (acquired by Synopsys) in the early 2000s. He earned his Ph.D. at KU Leuven University under the supervision of Georges Gielen, and is now currently co-founder and CTO of Solido Design Automation.

I love Trent's personal story as well. He grew up on a pig farm in Saskatchewan. As of this writing, Saskatchewan is 251,700 square miles with a population of just over 1 million souls. In contrast, California is 163,696 square miles with a population of 37 million souls. And Saskatchewan gets *cold*.

Trent once told me a story about farm life. When the weather gets to about $-40°$ (it is the same in Celsius or Fahrenheit), it freezes the valves for the pigs' outdoor watering bowls. To prevent damage to the plumbing, he had to pour hot water over the valves to thaw the ice, then re-fasten some tiny bolts. This latter step required taking his gloves off. He had about 15 s to fasten the bolts and get his gloves back on before freezing his fingers (and risking frostbite). Trent is a fountain of amazing farm stories from his boyhood. I grew up in California, and the biggest obstacle I had to starting the day was deciding if I was going to wear a long or short sleeve shirt that day. These experiences surely had a huge influence in building Trent's character.

As I have come to know Trent, I have also learned that he has a broad range of interests, from neuroscience, to music, to art. He complements his wife, who is an art curator with world-class training (Sorbonne, Paris) and work experience (The Louvre, Paris). Trent is truly a unique and entertaining renaissance man in the Da Vinci tradition.

On the technical side, Trent has a unique ability to invent algorithms that solve real design challenges, but not stop there. He takes the algorithms past the stage of prototype software that solves academic problems, shepherding them into commercial software usable by real designers doing production circuit design. At Solido, Trent has worked closely with Jeff Dyck, Kristopher Breen, and Solido's product development team, to deliver industrial-scale variation solutions.

This book helps custom IC designers to address variation issues, in an easy-to-read and pragmatic fashion. I believe this book will become an invaluable resource to the custom IC designer facing variation challenges in his/her memory, standard cell, analog/RF, and custom digital designs.

Enjoy the read… and never miss a chance to talk or listen to Trent!

Los Gatos, CA, July 2012   Jim Hogan

Acknowledgments

This book is the culmination of work by the authors and their colleagues, stretching back nearly a decade, in building variation-aware tools for circuit designers. It is a distillation of successful and not-so-successful ideas, of lessons learned, all geared towards making *better designs despite variation issues.*

We would like to thank those who reviewed the book, and provided extensive and valuable suggestions: Drew Plant, Mark Smith, Tom Eeckelaert, Jim Hogan, and Gloria Nichols. Thanks to those who have helped to provide and prepare real-world design examples: Ting Ku, Hassan Sharghi, Joel Amzallag, Drew Plant, Jiandong Ge, Anthony Ho, and Roshan Thomas. Thanks to the rest of the team in Solido Design Automation, who have helped us build and deliver high-quality tools. Thanks to the Solido investors and Solido board, without whom this work would not have been possible.

Thanks to our users and our partners, who have helped us shape our tools and methodologies for real-world use.

Finally, thank you for picking up this book! We hope that you find it useful (and valuable) in your own work.

Solido Design Automation Inc. Trent McConaghy
Canada, July 2012 Kristopher Breen
Jeff Dyck
Amit Gupta

Contents

Chapter 1
Introduction

Variation Effects, Variation-Aware Flows

Abstract This chapter introduces the problems of variation with respect to custom integrated circuit design. It then describes several design flows, and how well they handle or fail to handle variation. The chapter concludes with an outline for the rest of the book, which covers methodologies and tools for handling variation in industrial design settings.

1.1 Introduction

Variation is an expensive problem. Failing to effectively design for variation can cause product delays, respins, and yield loss. These are serious issues that directly impact the revenues, profits, and ultimately, valuations of semiconductor companies and foundries alike. The costs of variation problems trickle down the whole supply chain in the form of product delays, inability to meet market demands, finished product quality issues, and loss of customer confidence. This adds up to a massive annual loss. Calculating this loss would be extremely difficult, as there are many factors that contribute to the true cost of variation. However, we are aware of cases where variation problems have led to single product losses in excess of $100 million, so given the large number of semiconductor products available, it is intuitive that the annual losses due to variation are easily in the billions of dollars.

To make matters more challenging, physics and economics continue to co-conspire to drive us toward smaller process geometries. As transistors get smaller, performance targets increase, and supply voltages decrease, all making variation effects more pronounced. The variation problem continues to get worse, and the need to combat it with effective variation-aware design continues to become more essential.

Designing for variation is also expensive. Collecting data using test chips and building accurate models of variation is an intensive and complex procedure for

T. McConaghy et al., *Variation-Aware Design of Custom Integrated Circuits:*
A Hands-on Field Guide, DOI: 10.1007/978-1-4614-2269-3_1,
© Springer Science+Business Media New York 2013

foundries and design companies alike. Putting variation models to even basic use requires CAD tool investment, compute cluster investment, and CAD and IT support staff.[1] To effectively design for variation requires considerable designer expertise, additional time for careful analysis and design iterations, and time to perform thorough verification across the range of variation effects. Reducing variation effects at the design stage is a critical component to solving the overall variation problem, and this is best done with thoughtful design methodologies that are both rigorous and that can be completed within production timelines.

This book is for designers. It is a *field guide* to variation-aware design, outlining fast and accurate methods for eliminating risks and costs associated with variation. It is for people who design RF, analog, I/O, custom digital, digital standard cell, memory, automotive, or medical blocks or systems. That is, it is for people who work with schematics and SPICE simulators, rather than just RTL and related system code.[2] No revolution is necessary; this book describes minimal but specifically targeted extensions to existing corner-based, SPICE-based design methodologies. Furthermore, this book does not ask the reader to learn or re-learn deep statistical concepts or advanced algorithmic techniques; though for the interested reader, it does make those available in appendices.

In short, this book is about developing custom circuit designs that *work*, despite the effects of variation, and doing so within production timelines.

1.2 Key Variation Concepts

We begin by presenting an overview of key variation-aware design concepts: types of variables, types of variation, and terminology.

1.2.1 Types of Variables

Two types of variables affect a circuit's behavior:

- *Design variables (controllable variables):* These can be set by the designer, and together their choice constitutes the final design. These include the choices of topology, device sizes, placement, routing, and packaging.
- *Variation variables (uncontrollable variables):* These cannot be set by the designer in the final design; they happen due to various mechanisms that the designer cannot control. However, their values *can* be set *during design*, to

---

[1] CAD = computer-aided design, IT = information technology.

[2] RF = radio frequency, I/O = input/output, SPICE = Simulation Program with Integrated Circuit Emphasis (Nagel and Pederson 1973), and RTL = Resistor-Transistor Logic.

predict their effect on the design's performance. This ability to predict is the key enabler of variation-aware design.

1.2.2 Types of Variation

In integrated circuits, the variation variables may take many forms, which we now review.

Environmental variation: These variables include temperature, power supply voltage, and loads. In general, environmental variables affect the performance of the design once the circuit is operating in the end user environment. The design must meet target performance values across all pre-set environmental conditions; said another way, the worst-case performances across environmental corners must meet specifications. These pre-set conditions may be different for different circuits; for example, military-spec circuits typically must handle more extreme temperatures.

Modelset-based global process variation: These are die-to-die or wafer-to-wafer variations introduced during manufacturing, by random dopant fluctuations (RDFs) and more. Global process variation assumes that the variations affect each device in a given circuit (die) in an identical fashion. These variations affect device performance, for instance v_{th}, g_m, delay, or power, which in turn affect circuit performance and yield.

Traditionally, modelsets are used to account for global process variation. In modelsets, each NMOS model and each PMOS[3] model has a fast (**F**), typical (**T**), and slow (**S**) version, supplied by the foundry in netlist form as part of the Process Design Kit (PDK). The foundry typically determines the models by Monte Carlo (MC) sampling the device, measuring the mean and standard deviation of delay, then picking the sample with delay value closest to *mean* -3 * *stddev* (for **F** modelset), closest to *mean* (for **T** modelset), and closest *to mean* $+3$ * *stddev* (for **S** modelset).

The modelset approach to global process variation has traditionally been quite effective for digital design: **F** and **S** conservatively bounded the high and low limits of circuit speed; and since speed is inversely proportional to power, **F** and **S** indirectly bracketed power. The device-level performance measures of speed and power directly translated to the key digital circuit-level measures of speed and power. However, the modelset approach has not been adequate for analog and other custom circuits since modelsets do not bracket other performances such as slew rate, power supply rejection ratio, etc. When possible, designers have

---

[3] NMOS = N-channel MOSFET, PMOS = P-channel MOSFET, MOSFET = metal-oxide-semiconductor field-effect transistor.

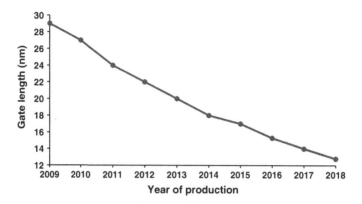

Fig. 1.1 Transistor gate length is shrinking (ITRS 2011)

compensated using differential topology designs; and when not possible, they used the modelsets anyway and hoped for the best.

Statistical global and local process variation: Whereas in the past, the modelset approach to handling global process variation was adequate for most cases, the situation is now changing. This is because gate lengths continue to shrink over time, as Fig 1.1 shows. This phenomenon—Moore's Law—is still happening: devices will continue shrinking in the foreseeable future. While transistors keep shrinking, *atoms* do not. For earlier technology generations, a few atoms out of place due to random dopant fluctuations or other variations did not have a major impact on device performance. Now, the same small fluctuations matter. For example, typically the oxide layer of a gate is just a few atoms thick, so even a single atom out of place can change device performance considerably. Statistical models can capture these variations. Local variation occurs within a single die, while global variation occurs across dies or wafers.

On modern process nodes, such as TSMC 28 nm or GF 28 nm, statistical models of variation are supplied by the foundry as part of the PDK. Larger semiconductor companies typically verify and tune these models with in-house model teams. A statistical model typically specifies the global random variables, the local random variables, and the distribution of those random variables. Another approach is to use modelsets for global process variation, and a statistical model for local variation only.

There are many approaches to modeling statistical variation. Probably the best-known approach is the Pelgrom mismatch model (Pelgrom and Duinmaijer 1989). In this model, matched devices are identified beforehand, such as devices in a current mirror, and the variance in threshold voltage V_t between matched devices is estimated. The theory is based on simple hand-based equations for transistors in the saturation region, which makes them poorly suited for calibration from tester-gathered MOS data, or for other transistor operating regions.

Since Pelgrom's famous work, many improved models have emerged. An example is the Back-Propagation-of-Variance (BPV) statistical model (Drennan

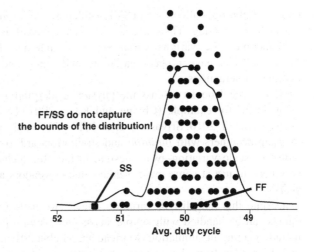

Fig. 1.2 FF/SS Corners versus Distribution, for the average duty cycle output of a phase-locked loop (PLL) voltage-controlled oscillator (VCO), on GF 28 nm. Adapted from (Yao et al. 2012)

and McAndrew 2003). It does not require specification of mismatch pairs. It is more accurate because it directly models the underlying physical variables as independent random variables (e.g. oxide thickness, substrate doping concentration) which can be readily calibrated by silicon measures. It can also account for both global and local variation. Beyond BPV, research on more accurate models continues, and foundries will continue to incorporate them into PDKs.

Figure 1.2 compares FF/SS variation versus statistical variation on a GF 28 nm process for a performance output of a typical analog circuit. We see that the FF/SS modelset does not adequately capture the performance bounds of the circuit, reconfirming our claim that FF/SS corners are not adequate on modern geometries for many types of custom circuits.

Layout parasitics: These resistances and capacitances (RCs) are not part of the up-front design, but rather emerge in the silicon implementation. They form within devices, between devices and interconnect, between devices and substrate, between interconnects, and between interconnect and substrate. Their effects are most concerning in circuits operating at higher frequencies (e.g. RF), or lower power supply voltages which have less margin. The challenge with layout parasitics is that one needs the layout to measure them, yet they affect electrical performance, which needs to be handled during front-end design, the step *before* layout. At advanced process nodes (e.g. 20 nm) where double patterning lithography (DPL) is used, the parasitics between layers can significantly impact performance.

Other types of variation: There are even more types of variation. Transistor aging/reliability includes hot carrier injection (HCI) and negative bias temperature instability (NBTI), which have been noted for some time, but are now becoming more significant. New aging issues include positive bias temperature instability (PBTI) and soft breakdown (SBD). There is electromigration (EM), which is aging on wires. There are layout-dependent effects (LDEs), which include well proximity effects (WPEs) and stress/strain effects. Thermal effects are becoming an issue with through-silicon-via (TSV)-enabled 3D ICs, which have less opportunity for air-based cooling. There are noise issues, crosstalk issues, and more.

Sometimes the various variation effects interact. For example, parasitics themselves can have process variation, and aging and process variation have nonlinear interactions.

Many recent books discuss the physical underpinnings of variation effects in great detail, for example (Chiang and Kawa 2007; Kundu and Sreedhar 2010; Orshansky et al. 2010; Srivastava et al. 2010). *This* book is complementary. It aims to equip designers with *intuition* and straightforward-to-apply *methodologies* for industrial-scale variation-aware design. In fact, the techniques in this book have now been in use at some major semiconductor vendors and foundries for several years.

Despite this long list of variation effects, we have found that many of these effects can be handled with simple steps, such as with post-layout simulation, or including aging in simulation (Maricau and Gielen 2010). Other types of variation may be hidden from the designer, for instance, using design rules, or with tools such as optical proximity correction.

In our experience, global process variation, local process variation, and environmental variation must be managed more directly by the designer, because (1) the effect on performance and yield is too large to be ignored, (2) they cannot be simply revealed by a single simulation, and (3) as we will see, simplistic sets of simulations such as comprehensive PVT corner analysis or thorough Monte Carlo analysis are too simulation-intensive.

This book focuses on global and local process variation, and environmental variation, with knowledge that many of the other effects are being adequately addressed orthogonally via appropriate tools and problem setup.

1.2.3 Key Variation-Related Terms

PVT variation is a combination of modelset-based global process variation (P) and environmental variation, including power supply voltage (V), temperature (T), load conditions, and power settings (e.g. standby, active).

Corners: A corner is a *point* in variation space. For example, a traditional PVT corner had a modelset value, a voltage value, and a temperature value, such as $\{modelset = FF, v_{dd} = 1.3\ V, T = 15\ °C\}$. The concept generalizes to include other types of variation. For example, a corner may have a value for each local process variable, such as $\{modelset = FF, v_{dd} = 1.3\ V, T = 15\ °C, M1\_Nsub = 0.23, M1\_tox = 0.12, M2\_Nsub = 0.21,...\}$. Due to DPL, RC parasitics are often modeled as corners too. As we will see, this generalized concept of corners is crucial to pragmatic variation-aware design.

Yield is the percentage of manufactured circuits that meet specs across all environmental conditions, expressed as a percentage, such as 95 %. Yield may also be expressed in alternate units of probability of failure, sigma1, and sigma2.

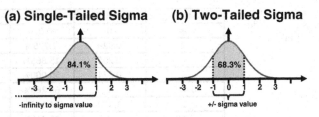

Fig. 1.3 Converting between yield and sigma. **a** Single-tailed sigma. **b** Two-tailed sigma

Probability of failure (p_{fail}) is another unit for yield, defined as $p_{fail} = 1 - yield(\%)/100$. For example, p_{fail} is 0.05 when yield is 95 %.

Sigma is a unit of yield, and can use either a single-tailed or two-tailed assumption, referred to as sigma1 and sigma2, respectively. Sigma1 yield is the area under a Gaussian curve from $-\infty$ to $+sigma$. Sigma2 yield is the area under the curve between $-sigma$ and $+sigma$. Figure 1.3 illustrates the difference between sigma1 and sigma2.

Figure 1.4 shows typical conversions among sigma1, sigma2, yield, and probability of failure.

High-sigma circuits: For an overall chip to have a reasonable yield (2–3 sigma), replicated blocks like standard cells and memory bitcells need to have much higher yields (4–6 sigma). The need to analyze and design such "high-sigma" circuits introduces qualitatively new challenges compared to 3-sigma design.

1.3 Status Quo Design Flows

We now review typical status-quo flows for designing custom circuit blocks. Figure (1.5a) shows the simplest possible flow.

- In the first step, the designer selects a topology.
- In the next step, he does initial sizing by computing the widths, lengths, and biases, typically from first principles equations against target power budget and performance constraints.
- In the third step, he makes modifications to the circuit sizing to improve performance. This step typically involves sensitivity analysis, sweeps, and other design exploration techniques, getting feedback from SPICE simulations.
- Starting from the sized schematic or netlist, the designer then does layout: device generation, placement, and routing.

At the end of this flow, the block is ready for integration into larger systems. The flow also works for system-level designs and higher by applying behavioral models or FastMOS/Analog FastSPICE simulators.

Fig. 1.4 Typical conversions among sigma1, sigma2, yield, and probability of failure

Single-tail Sigma	Two-tail Sigma	Yield	Prob. Failure
-∞	0	0.0%	1.000
-2	0.029	2.275%	.9772
-1	0.200	15.865%	.8414
0	0.674	50.000%	.5000
0.475	1	68.269%	.3173
1	1.410	84.134%	.1587
1.69	2	95.450%	.0455
2.78	3	99.730%	.0027
3.83	4	99.99366%	6.34e-5
4.86	5	99.9999427%	5.73e-7
5.88	6	99.9999998%	1.97e-9
+∞	+∞	100%	0.0

Fig. 1.5 Status quo design flows. **a** A simple flow. **b** Beginning to address variation with user-chosen PVT corners, ad-hoc statistical (Monte Carlo) sampling, and parasitic extraction

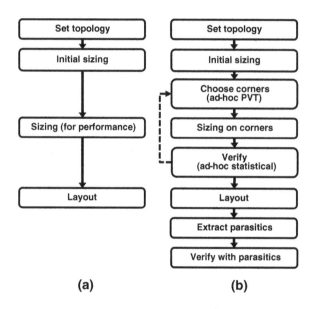

(a)   (b)

Of course, the flow of Fig. (1.5a) does not account for variations at all, leaving the final design highly exposed.

Figure (1.5b) shows an example status quo flow that begins to address variation. Starting with the simple flow of Fig. (1.5a), it adds user-chosen PVT corners, ad-hoc statistical Monte Carlo sampling, and post-layout parasitic extraction with SPICE-based verification.

While this is a big improvement in handling variation compared to the simple flow, it falls short in many regards:

- First, the user might not have chosen the PVT corners that bound worst-case performance, which means the design is optimistic and could fail in the field. Or, to be on the safe side, he used *all* the PVT corners. This resulting large number of corners means painfully long sizing iterations.

- Verifying statistical effects with a traditional Monte Carlo (MC) tool tends to be highly ad-hoc. How many MC samples should the designer choose? How does he measure whether the design "passes" or not? Furthermore, if he (somehow) decides that the design does not pass the statistical verification, how does he incorporate statistical effects into the sizing iterations? If he simply chose the MC samples that failed, those samples may have been too improbable (leading to an overly pessimistic design), or too probable (leading to an overly optimistic design). Furthermore, simply running MC is not adequate for high-sigma problems, which may require millions or billions of samples to verify a target failure rate.
- When verifying with parasitics on the layout-extracted netlist, the traditional flow ignores the effect of PVT and statistical variation. Conceivably, parasitics alone may not make the circuit fail; however, when combined with other variations, they could lead to failure.

In short, the status-quo flows are either slow or inaccurate with respect to PVT variations, slow or inaccurate with respect to statistical process variations, and inaccurate with respect to other variations.

1.4 A Fast, Accurate Variation-Aware Design Flow

Handling the many types of variation quickly yet accurately may seem like a daunting task. One key is to start with a nominal design, and incrementally add in the types of variation that may matter, at the appropriate times. The other key lies in fast, automated analysis technologies (for speed), using SPICE-in-the-loop and statistical confidence-based convergence (for accuracy).

Figure (1.6b) illustrates a fast, accurate variation-aware design flow. For easy comparison, Fig. (1.6a) is the status quo flow presented earlier. The changes from flow *a* to *b* are minimal yet precisely targeted, replacing ad-hoc steps with fast yet accurate variation-handling capabilities.

We now give details on the fast-yet-accurate variation-aware design flow given in Fig. (1.6b).

- The designer sets the topology and performs initial sizing in the usual fashion.
- After initial sizing, the designer runs a PVT corner extraction, which finds worst-case PVT corners. He can then design against these PVT corners until the specs are met across all corners. He may use sensitivity analysis, sweeps, and other design exploration tools to accomplish this. Once met, the designer runs a PVT verification, which finds worst-case PVT corners with confidence. A specialized tool performs PVT corner extraction and verification, quickly but accurately using SPICE-in-the-loop.
- [If appropriate] After PVT sizing, the designer runs a statistical corner extraction at a specified target sigma (e.g. 3-sigma, or a higher sigma value). He can then design against these statistical corners until the specs are met across all

Fig. 1.6 **a** The status quo ad-hoc variation-aware flow is slow and inaccurate. **b** A fast yet accurate variation aware flow leverages confidence-driven automated analyses

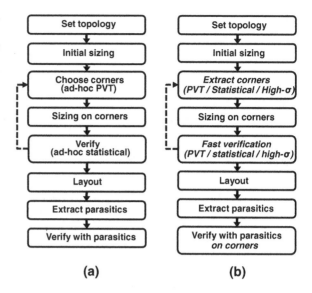

corners. Once met, he runs a statistical verification, which automatically determines with statistical confidence that the target sigma is met. Specialized tools perform statistical corner extraction and verification, quickly but accurately using SPICE-in-the-loop.

- [If appropriate] After layout, the designer simulates the parasitic-extracted netlist against the pre-layout PVT and/or statistical corners.

A variation issue is "appropriate" if the circuit may be susceptible to that issue, if adequate models exist to measure the issue, and (as a matter of unfortunate practicality) if the schedule permits.

The preceding discussion focused only on the main variation issues. Other variation effects may be inserted pragmatically as needed, as corners (like PVT or statistical variation), or as post-layout measures (like parasitics).

1.5 Conclusion/Book Outline

This chapter reviewed key variation concepts and related issues, and discussed status quo flows, and the issues with each flow. It then presented a pragmatic variation-aware flow that is fast yet accurate, shown in Fig. (1.6b). One key to the flow is to start with a nominal design, and incrementally add in the types of variation that may matter, at the appropriate times. The other key is fast, automated analysis technologies (for speed), using SPICE-in-the-loop and statistical confidence-based convergence (for accuracy).

The rest of this book focuses on the methodologies and tools to enable the fast-yet-accurate variation-aware flow. It has emphasis on techniques of corner extraction and verification, which are central to these methodologies.

The first several chapters are grouped according to the type of variation. There is a chapter for PVT-style variation, where global variation is accounted for using modelsets like *FF/SS*, in addition to voltage and temperature variation. There is a chapter for 3-sigma statistical variation, where global or local variation is modeled with a distribution. There is a chapter for high-sigma statistical variation. Each of these chapters discusses corner extraction and verification. We also include a chapter to help gain insight into distributions. Finally, there is a chapter on design: manual sizing, automated sizing, and the spectrum in between.

For the interested reader, deeper statistical concepts and advanced algorithmic techniques are made available in appendices.

The specific chapters are:

- Chapter 2: **PVT analysis:** This chapter surveys other PVT-based approaches and flows, then describes a novel "confidence-driven global optimization" technique for PVT corner extraction and verification. It concludes with two design examples from industry.
- Chapter 3: **Primer on probabilities:** This chapter takes a visually-oriented approach for insight about probability densities. It is useful standalone, when one is looking at histograms and normal quantile plots; and as a foundation to the next two statistically-oriented chapters.
- Chapter 4: **Three-sigma statistical analysis:** This chapter reviews other approaches and flows for three-sigma analysis, describes novel "density estimation and Optimal Spread Sampling" techniques for 3-sigma statistical corner extraction and verification, and concludes with industrial case studies.
- Chapter 5: **High-sigma statistical analysis:** This chapter surveys other high-sigma approaches and flows, then describes a novel technique for high-sigma statistical corner extraction and verification. It discusses full-PDF extraction and system-level analysis, and concludes with several industrial case studies.
- Chapter 6: **Variation-aware design:** Whereas the previous chapters focused on *analysis* techniques for corner extraction and verification, this chapter complements them with a focus on *design* techniques. It discusses three complementary approaches to sizing: manual, automated, and a new idea that integrates manual and automated approaches.
- Chapter 7: **Conclusion:** This chapter wraps up the book.

References

Chiang C, Kawa J (2007) Design for manufacturability and yield for nano-scale CMOS. Springer, Dordrecht

Drennan PG, McAndrew CC (2003) Understanding MOSFET mismatch for analog design. IEEE J Solid State Circuits 38(3):450–456

ITRS (2011) International Technology Roadmap for Semiconductors, http://itrs.net

Kundu S, Sreedhar A (2010) Nanoscale CMOS VLSI circuits: design for manufacturability. McGraw-Hill, NY

Maricau E, Gielen GGE (2010) Efficient variability-aware NBTI and hot carrier circuit reliability analysis. IEEE Trans CAD Integr Circuits Syst 29(12):1884–1893

Nagel LW, Pederson DO (1973) SPICE (Simulation Program with Integrated Circuit Emphasis). Memorandum No. ERL-M382. University of California, Berkeley, Apr 1973

Orshansky M, Nassif S, Boning D (2010) Design for manufacturability and statistical design: a constructive approach. Springer, NY

Pelgrom MJ, Duinmaijer ACJ (1989) Matching properties of MOS transistors. IEEE J Solid State Circuits 24:1433–1440

Srivastava A, Sylvester D, Blaauw D (2010) Statistical analysis and optimization for VLSI: timing and power. Springer, NY

Yao P et al (2012) Understanding and designing for variation in GLOBALFOUNDRIES 28 nm technology. In: Proceedings of design automation conference (DAC), San Francisco

Chapter 2
Fast PVT Verification and Design

Efficiently Managing Process-Voltage-Temperature Corners

Abstract This chapter explores how to design circuits under PVT variation effects, as opposed to statistical process variation effects. Process, voltage, and temperature (PVT) variations are taken into account by individually varying P, V, and T over their allowable ranges and analyzing the subsequent combinations or so-called PVT corners. In modern designs, there can be hundreds or thousands of PVT corners. This chapter reviews design flows to handle PVT variations, and compares them in terms of relative speed and accuracy. It introduces a "Fast PVT" flow and shows how that flow has excellent speed and accuracy characteristics. It describes the Fast PVT algorithm, which is designed to quickly extract the most relevant PVT corners. These corners can be used within a fast and accurate iterative design loop. Furthermore, Fast PVT reliably verifies designs, on average 5x faster than the method of testing all corners on a suite of benchmark circuits. This chapter concludes with design examples based on the production use of Fast PVT technology by industrial circuit designers.

2.1 Introduction

PVT effects are modeled as a set of corners. A PVT corner has a value for each PVT variable. For example, a single PVT corner might have a modelset value of FF, a supply voltage of 1.2 V, a temperature of 25 °C, and power setting of *standby*. A set of PVT corners is created by enumerating the complete combinatorial set of variations. For example, if there are 5 values of modelset, 3 values of voltage, 3 values of temperature, and 4 power modes, there would be a total of $5 \times 3 \times 3 \times 4 = 180$ combinations in the PVT corner set.

There are some cases in custom design where a PVT approach to variation may be used instead of a statistical approach:

T. McConaghy et al., *Variation-Aware Design of Custom Integrated Circuits:*
A Hands-on Field Guide, DOI: 10.1007/978-1-4614-2269-3_2,
© Springer Science+Business Media New York 2013

- Local process variation (mismatch) has negligible effect on the circuit's performance, and modelset corners (FF/SS) accurately bound global process variation's effect on circuit performance. This is often the case with digital standard cells, especially for older process technologies.
- Sufficiently accurate models of statistical process variation are not available. This is typically the case for older process technologies.
- PVT corners have tolerable accuracy in bounding performance, and statistical analysis is deemed too simulation-intensive to complete. For example, many signoff flows mandate a thorough PVT analysis.
- Local process variation affects performance or modelset corners do not bound performance, statistical models are available, but there is no clear statistically-based design methodology. This is sometimes the case with analog design: designers know that FF/SS corners are inaccurate, but do not have a fast statistically-based design flow. If this is the case, a PVT-based flow is *not* appropriate. This book is here to help: the chapters that follow this chapter describe flows to do fast-yet-accurate *statistically*-aware design.

The aim in PVT analysis is to find the worst-case performance values across all PVT corners, and the associated PVT corner that gives the worst-case performance. This is done for each output. In PVT-aware design, the aim is to find a design that maximizes performance or meets specifications across all PVT corners.

Traditionally, it has only been necessary to simulate a handful of corners to achieve coverage: with FF and SS process (P) corners, plus extreme values for voltage (V) and temperature (T), all combinations would mean $2^3 = 8$ possible corners.

With modern process nodes, many more process corners are often needed to properly bracket process variation across different device types. Here, "to bracket" means to find two corners for each spec, one that returns the maximum value and another that returns the minimum value of the respective performance output. Furthermore, transistors are smaller, performance margins are smaller, voltages are lower, and there may be multiple supply voltages. To bracket these variations, more variables with more values per variable are needed.

Adding more PVT variables, or more values per variable, quickly leads to a large number of corners. Even a basic analysis with 4 device types (e.g. NMOS, PMOS, resistor, capacitor) and 4 other variables (e.g. temperature, voltage, bias, load) with 3 values each results in $3^{(4+4)} = 6,561$ corners.

The problem gets worse at advanced nodes that have double patterning lithography (DPL), where the RC parasitics among the masks degrade performance. To account for this, a tactic is to treat the bounds on RC parasitic variation as corners by extracting different netlists that represent each possible extreme. This results in a 10–15x increase in the number of corners.

Power verification needs corners for each power mode (e.g. quick boot, cruising, read, write, turbo, and standby), which also increases the total number of corners.

The problem is that simulating each corner can take several seconds, minutes, or even hours for longer analyses. To simulate all possible corners could take hours or even days, which is too time-consuming for most development schedules. Designers may try to cope with this limitation by guessing which corners cause the worst-case performance, but that approach is risky: a wrong guess could mean that the design has not accounted for the true worst-case, leading to a failure in testing followed by a re-spin, or worse, failure in the field.

The technique of finding the worst-case PVT corners is most effectively employed in the context of the design loop, where the engineer is changing circuit design variables such as transistor width (W) and length (L) in order to find the design with the best performance under the worst-case PVT conditions.

What is therefore required is a rapid, reliable methodology to quickly identify the worst-case PVT corners when there are hundreds, thousands, or even tens of thousands of possible corners. We define these attributes as follows:

- *Rapid*: Runs fast enough to facilitate both iterative design and verification within production timelines.
- *Reliable*: Finds the worst-case corners with high confidence.

These attributes must be met in the context of a practical flow for PVT-aware design.

This chapter first reviews various design flows to handle PVT variation, including a flow using a Fast PVT corner approach. Next, it considers possible algorithms to implement a Fast PVT corner approach, including a global optimization-based approach that delivers on the rapid and reliable attributes. Finally, it presents results of the chosen approach on real-world production designs. This chapter's appendix provides more details on the global optimization-based Fast PVT approach.

2.2 Review of Flows to Handle PVT Variation

This subsection reviews various flows for handling PVT variation that a designer might consider. These flows include simulating all combinations, guessing the worst-case PVT corners, and a new approach that uses Fast PVT capabilities. This section compares the flows in terms of speed and accuracy.

To illustrate each flow, we provide an estimated number of simulations and design time for a representative real-world circuit. The real-world circuit is the VCO of a PLL on a 28 nm TSMC process technology. It has two output performance measures with specifications of: $48.3 <$ duty cycle < 51.7 %, and $3 <$ Gain < 4.4 GHz/V. Its variables are *temperature*, $V_{ah,vdd}$, $V_{a,vdd}$, $V_{d,vdd}$, and model set (any one of 15 possible sets). All combinations of all values of variables leads to 3375 corners. Since there are two output performance measures, and each output has a lower and an upper bound, there are up to 4 PVT corners that cause worst-case performances. We used a popular commercial simulator.

Fig. 2.1 PVT flow: full factorial

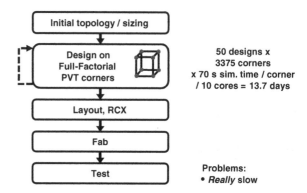

Initial topology / sizing

Design on Full-Factorial PVT corners

50 designs x
3375 corners
x 70 s sim. time / corner
/ 10 cores = 13.7 days

Layout, RCX

Fab

Test

Problems:
• *Really* slow

Because the flows include changing the design variables, we need a way to compare different approaches in a fair fashion, independent of designer skill and level of knowledge the designer has about the circuit. To do this, we use a simple assumption that in the design loop, the designer considers 50 designs. In our comparisons, we consider time spent with simulations. We do not consider the time spent modifying the design. Also, since multi-core and multi-machine parallel processing is commonplace, for each approach we assume that there are 10 cores running in parallel as our computing resource.

2.2.1 PVT Flow: Full Factorial

Full factorial is the simplest of all flows, shown in Fig. 2.1. In this flow, the designer simply simulates all possible PVT corners at each design iteration. It is comprehensive, and therefore perfectly accurate, to the extent that PVT variation is an accurate model of variation. However, since each of 50 design iterations takes 3375 simulations, it is very slow, requiring 13.7 days of simulation time even when using 10 parallel cores.

2.2.2 PVT Flow: Guess Worst-Case

Figure 2.2 illustrates this flow. After the topology and initial sizing are chosen, the designer uses expertise and experience to guess which PVT corners are likely to cause worst-case performances, without any simulations. Then, the designer simply designs against those corners.

The advantage of this approach is its speed, as it requires no simulations to select corners, and each design iteration only requires simulating the selected corners, which by the designer's reckoning represent respectively the upper and lower specification for each of two outputs. The disadvantages of this method are

Fig. 2.2 PVT flow: guess
worst-case

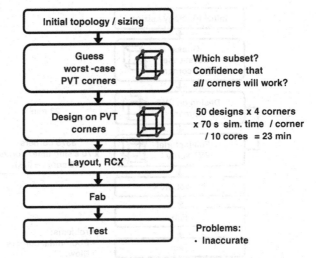

Initial topology / sizing

Guess
worst -case
PVT corners

Which subset?
Confidence that
all corners will work?

Design on PVT
corners

50 designs x 4 corners
x 70 s sim. time / corner
/ 10 cores = 23 min

Layout, RCX

Fab

Test

Problems:
· Inaccurate

poor accuracy and reliability. If the designer's worst-case PVT guess corner is
wrong, then the known worst-case performance is optimistic. This can be a big
issue. For example in power verification, it could mean that the circuit may not
meet the required power budget, which in turn translates into poor battery life on
the mobile device it is built into.

2.2.3 PVT Flow: Guess Worst-Case + Full Verification

Figure 2.3 shows this flow. It is similar to the previous flow, but adds a step of
running all combinations of PVT corners (full factorial) after the design step. This
overcomes a key oversight of the previous flow, ensuring that the circuit is verified
to the target specs. However, it is possible that the additional verification step finds
new worst-case corners that make the design fail specs. To pass at these new PVT
corners, the design must be improved at the new corners and verified again. This
requires more simulations.

Overall, the flow is more accurate than before, but the full factorial steps are
quite slow, resulting in a long overall runtime (14.2 h on our example circuit).
Also, if extra design iterations are needed, it could also add substantially to the
overall design cycle time, and the long full verification would need to be repeated.

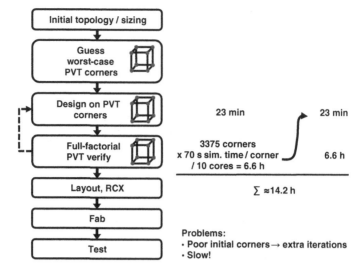

Fig. 2.3 PVT flow: guess worst-case + full verification

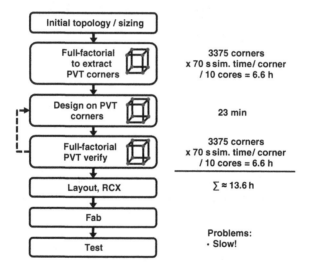

Fig. 2.4 PVT flow: good initial corners + full verification

2.2.4 PVT Flow: Good Initial Corners + Full Verification

Figure 2.4 illustrates this flow. This flow addresses the issue of the previous flow, which is that incorrectly selecting worst-case corners initially can lead to additional design and verification iterations.

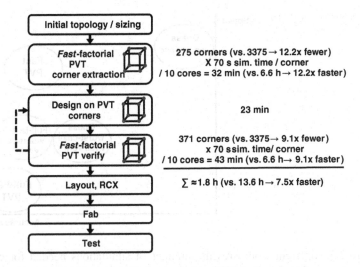

Fig. 2.5 PVT flow: fast factorial PVT (or simply Fast PVT)

The idea with this flow is to simulate all corner combinations at the beginning to identify the correct worst-case PVT corners. Then, the user designs against these corners. Then, in case there are strong interactions between design variables and PVT variables, the user runs full factorial PVT. Usually, PVT corners identified in the beginning will continue to be the worst-case corners; if not, the user designs against the new worst-case corners and verifies again.

Overall, the flow has a comparable runtime and accuracy to the previous flow, but typically only needs one design round because PVT design is done using verified worst-case corners.

This flow is accurate, but still fairly slow because it involves running full factorial PVT twice. On our example circuit, this would still require 13.6 h of simulation time.

2.2.5 PVT Flow: Fast PVT

Figure 2.5 shows this flow. It includes a fast way to handle the two most expensive steps: PVT corner extraction, and PVT verification. It is like the previous flow, but replaces the full factorial PVT steps with *fast factorial* PVT steps.

The flow is as follows. First, the designer completed the initial topology selection and sizing. Then, the designer invokes a fast factorial PVT tool, which we will refer as Fast PVT. Fast PVT efficiently extracts good worst-case PVT corners. The user then designs against these PVT corners. To cover the case where there are strong interactions, he invokes the Fast PVT tool again, verifying that the worst-case PVT corners are found. If needed, he does a second round of design and verification.

Fig. 2.6 Summary of PVT flows

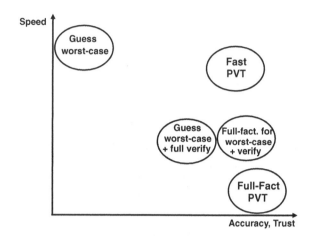

Figure 2.5, top right, compares the number of simulations needed for corner extraction using full factorial PVT against the number needed using Fast PVT, where Fast PVT uses an algorithm described later in this chapter. Fast PVT finds exactly the same worst-case corners as are found using full factorial PVT. We see that Fast PVT needs 12.2x fewer simulations, and runtime is reduced from 6.6 h to 32 min.

Figure 2.5, center right, compares doing PVT final verification using full factorial with using Fast PVT. Fast PVT again finds exactly the same worst-case corners as full factorial PVT. We see that Fast PVT needs 9.1x fewer simulations, and that runtime is reduced from 6.6 h to 43 min.

The time taken by design iterations is now comparable to the time taken by PVT corner extraction or PVT verification.

In this circuit example, the overall runtime of a Fast PVT-based flow is 1.8 h, which is about 7.5x faster than the 13.6 h required for the previous flow.

We will see later in this chapter how these results are illustrative of Fast PVT's behavior: it will nearly always find the worst-case corners with about 5x fewer simulations than a full-factorial PVT approach.

Subsequent sections of this chapter will review how Fast PVT can be implemented.

2.2.6 Summary of PVT Flows

Figure 2.6 summarizes the different PVT flows described above, comparing them in terms of speed and accuracy. Guessing the worst-case PVT corners without any further verification is the fastest flow, but heavily compromises accuracy. On the flipside, full factorial PVT in the design loop is very accurate but very slow. There are hybrid variants employing full factorial final verification, but replacing full factorial in the loop with corner extraction using initial guessing or full factorial.

However, the full factorial characteristic slows these approaches down. The Fast PVT approach, in the top right, gives the best tradeoff between speed and accuracy, by using a fast factorial approach to extracting corners and to verifying designs across PVT variation.

2.3 Approaches for Fast PVT Corner Extraction and Verification

Fast PVT aims to find the worst-case PVT corners from hundreds or thousands of candidate corners, using as few simulations as possible yet maximizing the designer's chances of finding the truly worst PVT corners. To be suitable for production design, the approach needs to be both rapid and reliable. Engineers and researchers have investigated a number of approaches to do Fast PVT corner extraction. This section summarizes some of the popular approaches and highlights challenges with each method.

2.3.1 Full Factorial

Running all combinations is not fast, but it serves as a baseline for comparison for other methods. As discussed, this can be very time-consuming, taking hours or days. On the positive side, it always returns the true worst-case corners.

2.3.2 Designer Best Guess

Guessing may not produce reliably accurate results, but it serves as a baseline. Here, the designer makes guesses based on experience about what the worst-case corners may be. The advantage of this approach is speed, as guessing requires no simulations. The disadvantage is lack of reliability; a wrong guess can mean failure in testing or in the field. Reliability is strongly dependent on the designer's skill level, familiarity with the design, familiarity with the process, and whether the designer has adequate time to make a qualified guess. In practice, it is difficult to consistently meet all of these goals, which makes a guessing-based approach inherently risky.

2.3.3 Sensitivity Analysis (Orthogonal Sampling, Linear Model)

In this approach, each variable is perturbed one-at-a-time, and the circuit is simulated for each variation. An overall linear response surface model (RSM) is constructed. The worst-case PVT corners are chosen as the ones predicted to give worst-case output values by the linear model. This method is fast, as it only requires $n + 1$ simulations for n variables (the +1 is for typical case). However, reliability is poor: it can easily miss the worst-case corners because it assumes a linear response from PVT variables to output, and assumes that there are no interactions between variables; this is often not the case.

2.3.4 Quadratic Model (Traditional DOE)

In this approach, the first step is to draw $n \times (n\text{-}1)/2$ samples in PVT space using a fractional factorial design of experiments (DOE) (Montgomery 2004), then simulate them. The next step constructs a quadratic response surface model (RSM) from this input/output data. Finally, the worst-case PVT corners are the ones that the quadratic model predicts as worst-case output values. While this approach takes more simulations than the linear approach, it is still relatively fast because the number of input PVT variables n is relatively small. However, reliability may still be poor because circuits may have mappings from PVT variables to output that are more nonlinear than simple quadratic.

2.3.5 Cast as Global Optimization Problem (Fast PVT)

The idea here is to cast PVT corner extraction and PVT verification as an optimization problem: search through the space of candidate PVT corners x, minimizing or maximizing the simulated performance output value $f(x)$. Under that problem specification, the aim is to solve the optimization problem reliably, with as few simulations as possible. The optimization must be *global*: working independently of any initial PVT corner values, and it must not get stuck in local optima. This is a promising idea because it directly addresses the designer task of finding the worst-case PVT corner, and if implemented well, delivers both good accuracy and speed.

Figure 2.7 illustrates how PVT verification can be cast as a global optimization problem. The x-axis represents the possible PVT values, which in this case is just the temperature variable. The y-axis is the performance metric to maximize or minimize, which in this case is the goal to maximize power. The curve is the response of the circuit's power to temperature, found via SPICE simulation. The objective in optimization is to try different x-values, measuring the y-value, and

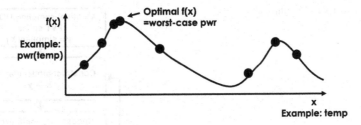

Fig. 2.7 PVT verification cast as a global optimization problem

using feedback to try more x-values, ultimately maximizing the y-value. In this case, different values of temperature are being selected and simulated to find the maximum value of power. The top of the hill in Fig. 2.7 right is a local optimum, as none of the nearby x-values have a higher y-value. We do not want the *Fast PVT* algorithm to get stuck in this local optimum; it should instead find the top of the hill in Fig. 2.7 left, which is the global optimum. In other words, over all possible values of temperature, *Fast PVT* should give the worst-case output performance (in this case, maximum power).

With PVT verification recast as a global optimization problem, we can now consider various global optimization approaches to solve it. Global optimization is an active research area with a long history, spanning techniques with labels like "Branch & Bound" (Land and Doig 1960), "Multi-Coordinate search" (Huyer and Neumaier 1999), and "Model-Building Optimization" (MBO) (Jones et al. 1998). We focus our energy on MBO because it is rapid, reliable, and easy for users to understand. Subsequent sections will validate these claims and describe the approach in general.

In this work, by "Fast PVT", we refer specifically to the solution to the general challenge of finding worst-case corners accurately and efficiently. Fast PVT is then the approach that casts the problem as a global optimization problem, and uses the MBO-based approach to solving the global optimization problem.

2.4 Fast PVT Method

2.4.1 Overview of Fast PVT Approach

The overall approach is to cast PVT corner extraction and PVT verification as a global optimization problem, and solve it using Model-Building Optimization (MBO). MBO-based Fast PVT is rapid because it builds regression models that make maximum use of all simulations so far in order to choose the next round of simulations. MBO uses an advanced modeling approach called Gaussian Process Models (GPMs) (Cressie 1989). GPMs are arbitrarily nonlinear, making no assumptions about the mapping from PVT variables to outputs. Fast PVT is

Fig. 2.8 Fast PVT algorithm

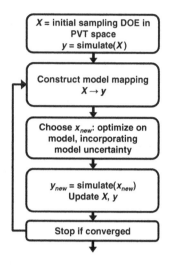

reliable because it finds the true worst-case corners, assuming appropriate stopping criteria. Fast PVT is user-friendly and easy to adopt because it lends itself well to visualization, and its algorithmic flow is easy for designers to understand.

As discussed, corner extraction and verification are actually two distinct tasks. Fast PVT has slightly different algorithms, depending on the task:

- *Corner extraction* is for finding PVT corners that the designer can subsequently design against. Corner extraction runs an initial design of experiments (DOE), predicts all values using advanced modeling, then simulates the predicted worst-case predicted corners for each output. There is no adaptive component.
- *Verification* keeps running where corner extraction would have stopped. It loops to adaptively test candidate worst-case corners while updating the model and improving the predictions of worst-case. It stops when it is confident it has found the worst-case. Verification takes more simulations than corner extraction, but is more accurate in finding the worst-case corners.

Figure 2.8 shows the Fast PVT algorithm. Both Fast PVT corner extraction and verification start by drawing a set of initial samples, X (i.e. corners), then simulating them, y. A model mapping $X \rightarrow y$ is constructed.

After that, corner extraction simply returns the predicted worst-case points. Verification proceeds by iteratively choosing new samples via advanced modeling, and then simulating the new samples. It repeats the modeling/simulating loop until the model predicts, with 95 % confidence, that worse output values are no longer possible. When choosing new samples, it accounts for both the model's prediction of the worst-case, as well as the model's uncertainty (to account for model blind spots).

Fast PVT is rapid because it simulates just a small fraction of all possible PVT corners. It is reliable because it does not make assumptions about the mapping from PVT variables to outputs, and explicitly tracks modeling error. Later in this paper, results on several benchmark circuits will further validate the technique's speed and reliability.

For a detailed explanation of the steps in Fig. 2.8, and theoretical details, we refer the reader to Appendix A. For further details on the GPM modeling approach, we refer the reader to Appendix B.

2.5 Fast PVT Verification: Benchmark Results

2.5.1 Experimental Setup

This section catalogs Fast PVT verification benchmark results on a suite of problems: 13 circuits with a total of 118 outputs, based on industry applications and PVT settings. The circuits include a shift register, two-stage bucket charge pump, two-stage opamp, sense amp, second-order active filter, three-stage mux, switched-capacitor amplifier, active bias generator, buffer chain, and SRAM bit-cell. The number of candidate PVT corners ranges from 130 to 1800. All circuits have devices with reasonable sizings. The device models are from modern industrial processes ranging from 65 to 28 nm nodes.

We performed two complementary sets of benchmarking runs. The methodology for each set is as follows.

Per-circuit benchmarks methodology. First, we simulated all candidate PVT corners and recorded the worst-case value seen at each output; these form the suite of "golden" reference results. Then, we ran Fast PVT verification *once per circuit* (13 runs total) and recorded the worst-case values that were found for each output, and how many simulations Fast PVT took.

Per-output benchmarks methodology. We obtained the "golden" reference results by simulating all PVT corners. Then, we ran Fast PVT verification *once for each output of each circuit* (118 runs total) and recorded the worst-case values that were found for each output, and how many simulations Fast PVT took.

2.5.2 Experimental Results

In all the runs, Fast PVT successfully found the worst-case point. This is crucial: speedup is only meaningful if Fast PVT can find the same worst-case results as a full-factorial (all corners) PVT run.

Figure 2.9 shows the distribution of speedups using Fast PVT verification, on the per-circuit benchmarks (left plot), and per-output benchmarks (right plot). Each point in each plot is the result of a single run of Fast PVT. Speedup is the number of full factorial PVT corners, divided by the number of PVT corners that Fast PVT verification needs to find the worst case outputs and then to stop.

Fig. 2.9 PVT verification
results, showing speedup per
problem. Left: Speedup on
per-circuit benchmarks (≥ 1
outputs per circuit). Right:
Speedup on per-output
benchmarks

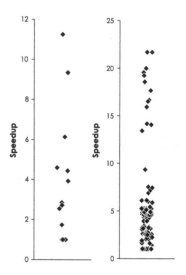

On the 13 per-circuit benchmark runs (Fig. 2.9 left), Fast PVT found the true
worst-case corners 100 % of the time. The average speedup is 4.0x, and the
maximum speedup is 11.0x.

On the 118 per-output benchmark runs (Fig. 2.9 right), Fast PVT found the true
worst-case corners 100 % of the time. The average speedup is 5.0x, and the
maximum speedup is 21.7x.

We found that Fast PVT verification speedups are consistently high when the
total number of corners is in the hundreds or more; for fewer corners, the speedup
can vary more. A case of 1.0x speedup (i.e. no speed-up) simply means that Fast
PVT did not have enough confidence in its models to complete prior to running all
possible corners.

2.5.3 Post-Layout Flows

Post-layout flows are worth mentioning. Figure 2.5 shows a PVT-aware flow for
designing prior to layout. This flow can be readily adapted to handle post-layout as
well. Because post-layout netlists tend to be more expensive to simulate, there are
options we can consider, in order of increasing accuracy and simulation time:

- Just simulate on a typical PVT corner. If specs fail, adjust the design to meet
  them, in layout space or sizing space, with feedback from the simulator on the
  corner. Finally, if time permits, re-loop with verification.
- Simulate with worst-case pre-layout PVT corners. If specs fail, adjust design as
  needed.
- Run Fast PVT verification. If specs fail, adjust design as needed.

As an example, we benchmarked Fast PVT on the PLL VCO circuit from Sect. 2.2 in its post-layout form. Fast PVT on the post-layout circuit took 171 simulations to find the true worst-case corners and to achieve verification-level confidence. Comparing this with the technique of simulating all 3375 combinations shows a 19.7x speedup.

2.5.4 Fast PVT for Multiple Outputs and Multiple Cores

The algorithm just described is for a single output, assuming a serial process. It is straightforward to extend the Fast PVT algorithm to handle more than 1 output, as follows. All outputs use the same initial samples. In each iteration, the algorithm for each output chooses a new sample. Model building for a given output always uses all the samples taken for all outputs.

It is also straightforward to parallelize the Fast PVT algorithm. Initial sampling sends off all simulation requests at once, and the iterative loop is modified slightly in order to keep available parallel simulation nodes busy.

2.5.5 Fast PVT Corner Extraction Limitations

Recall that the corner extraction task of Fast PVT runs the initial DOE, builds a model, predicts worst-case corners, and stops. It is not adaptive. Therefore, its main characteristics are that it is relatively fast and simple, but not as accurate as Fast PVT in verification mode in terms of finding the actual worst-case corners.

2.5.6 Fast PVT Verification Limitations

Stopping criteria and model accuracy govern the speed and reliability of Fast PVT verification. The most conservative stopping criteria would lead to all simulations being run, with no speedup compared to full factorial. On the other hand, stopping criteria that are too aggressive would result in stopping before the worst-case is found. Fast PVT verification strikes a balance, by stopping as soon as the model is confident it has found the worst-case. The limitation is that the model may be overly optimistic, for instance if it has missed a dramatically different region. To avoid suffering this limitation, Sect. 2.5.7 provides guidelines on measurements, since having the right measurements can result in improved model quality.

Model construction becomes too slow for >1000 simulations taken. In practice, if Fast PVT has not converged by 1000 simulations, it probably will not converge for more simulations, and will simply simulate the remaining corners. Appendix B has details.

Fast PVT performs best when there are fewer input PVT variables. For >20 variables, modeling becomes increasingly difficult and starts to require >1000 simulations. For these reasons, we do not recommend using Fast PVT with >20 input variables. Appendix B has details.

Fast PVT speedup compared to full factorial is dependent on number of candidate corners: the more candidate corners, the higher the speedup. This also means that if there is a small number of candidate corners (e.g. 50 or less), then the speedup is usually not significant (e.g. 2x, or even 1x).

In summary, while Fast PVT does have some limitations, it nonetheless provides significant benefit for production designs.

2.5.7 Guidelines on Measurements

Fast PVT's speed and accuracy depend on the accuracy of the model constructed, that is the model mapping from PVT input variables to outputs. The greater the model accuracy, the faster the algorithm convergence.

In designing measurements and choosing targets, the user should be aware of these guidelines:

- The outputs can be binary, and more generally, can have discontinuities. Outputs with these behaviors however typically require more simulations to model accurately.
- Some candidate samples can produce output measurement failures, as long as those measurement failures correspond to extreme values of one of the outputs being targeted.
- The outputs cannot contain simulator noise as the primary component of output variation; this situation results in random mappings from PVT variables to output. If this situation occurs, it usually means there is a problem with the measurement. A well-implemented Fast PVT algorithm should automatically detect random mappings from PVT variables to outputs.

2.6 Design Examples

This section presents two design examples based on production use of Fast PVT technology by industrial circuit designers.

Fig. 2.10 Folded-Cascode
amplifier with gain boosting

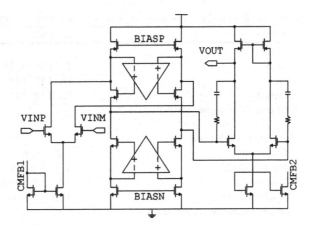

2.6.1 Corner-Based Design of a Folded-Cascode Amplifier

In this example, we examine the PVT corner analysis of an amplifier circuit. Figure 2.10 shows the amplifier schematic for this example.

Table 2.1 describes the required operating process and environmental conditions for the amplifier.

Performance outputs for the amplifier design include gain, bandwidth, phase margin, noise, power consumption, and area. PVT corner analysis and design are performed using Solido Variation Designer (Solido Design Automation 2012). Simulations are performed with the Cadence® Virtuoso® Spectre® Circuit Simulator (Cadence Design Systems 2012).

First, Fast PVT analysis is used in "corner extraction" mode to establish worst-case corners for each of the outputs of the design. The analysis needs to account for each output individually because the worst-case corners are rarely the same for all outputs. For this design, Fast PVT is configured to find worst-case corners corresponding to minimum gain, bandwidth, and phase margin, and maximum noise and power consumption.

From the 3645 combinations of process, voltage, temperature, bias, and load, Fast PVT corner extraction runs approximately 30 simulations to find a representative set of worst-case corners, resulting in a 122x simulation reduction for this first analysis step over full factorial simulation.

The worst-case corners found by Fast PVT corner extraction result in poor gain and phase margin performance for this design. Therefore, these corners are saved for use in the next step of the flow. Note that worst-case corners for all outputs are saved, not just those for poorly performing outputs. The corners for all outputs need to be included during design iteration in order to ensure that no outputs go out of specification under worst-case conditions when changes are made to the design.

Table 2.1 Required process and environmental conditions for amplifier design

Conditions	Values	Quantity
Process		
MOS	FF, FS, TT, SF, SS	5
Resistor	HI, TYP, LO	3
Capacitor	HI, TYP, LO	3
Total process combinations		45
Environmental conditions		
Temperature	−40, 27, 85	3
Supply voltage	1.45, 1.5, 1.55	3
Bias current[a]	9.5u, 10u, 10.5u	3
Load capacitance	160f, 170f, 180f	3
Total environmental Combinations		81
Total combinations (process and environmental)		3645

[a] Bias current variation affects the BIASN and BIASP voltages shown in the schematic

For the next step in the flow, the corners found during the initial Fast PVT corner extraction are used to examine the sensitivity of the design under worst-case conditions. Several opportunities are available for modifying the design to improve gain and phase margin under worst-case conditions, without trading off too much performance, power, or area. Such opportunities include adjusting the *W/L* ratios of the NMOS and PMOS bias transistors (the transistors with gates connected to BIASN and BIASP), as well as the sizing of the second stage differential pair and capacitors.

Once the sensitivity of the outputs is determined across the worst-case corners, the design is modified and the performance is checked by simulating at each worst-case corner. In this example, this iterative modify/simulate procedure is repeated four times to achieve satisfactory performance across all worst-case corners and to find an acceptable tradeoff between performance, power, and area.

After the design iterations are complete, Fast PVT verification is performed. The analysis confirms that design performance is acceptable across the entire range of PVT combinations. Fast PVT verification runs only 568 simulations to find the worst-case corners for this design out of the total 3645 combinations of process and environmental conditions.

Table 2.2 shows the number of simulations performed for each step in this flow. For comparison purposes, two other approaches were used on this design and are also summarized in the table. The first column summarizes the Fast PVT flow, the second column summarizes the flow of running all combinations and the third column uses the designer's best guess for determining worst-case corners. Although the "best guess" flow uses the least simulations, the resulting design does not perform well across all corners.

Table 2.2 Number of simulations required for Fast PVT design/verification flow

Step	Number of simulations (Fast PVT flow)	Number of simulations (full factorial flow)	Number of simulations (designer "best guess" flow)
Initial corner extraction	30	3,645	5
Design sensitivity across worst-case corners	115	115+[a]	115
Design iteration	20	20+	20
Verification	568	3,645	5
Total	733	7,425+	145[b]

[a] The full factorial flow simulation numbers assume that design sensitivity analysis and design iterations are performed with manually extracted corners from a full factorial PVT analysis. If design iterations are instead performed by running all corners, then the number of simulations would be even larger

[b] The best guess flow does not correctly identify the worst-case corners, and the resulting design performance is not satisfactory in that case

In summary, using the Fast PVT design and verification flow with the amplifier design reduces simulations, while achieving an overall robust design. Note that if there are failures during verification, further iterations are required, which can increase the number of simulations. However, the overall number of simulations in that case is still much less than running full factorial PVT, and in practice, this occurs relatively infrequently.

2.6.2 Low-Power Design

Low-power design is very important in mobile systems-on-a-chip (SoCs), due to the demand for longer battery life in smaller, lighter devices. Power consumption needs to be carefully analyzed across the different components in a mobile SoC to ensure that power consumption is acceptable and that it does not become too large under certain operating conditions.

A challenge in low-power design is that power consumption can vary significantly under different process and environmental conditions. Furthermore, the chip state has a dramatic impact on the power consumption. That chip state interacts with the process and environmental conditions, such that a state that causes little power consumption under one set of conditions may cause much more power consumption under a different set of conditions.

For this reason, it is important to simulate the design under many different process, environment, and state conditions. However, the number of combinations that need to be taken into account for each cell can be very large. This is especially problematic for circuits with long transient simulation times. The challenge is even greater when designing below 28 nm, where additional simulation of RC parasitic corners is required to capture interconnect variability. The analysis must then be repeated for each cell in the SoC and the results aggregated together. Finally, if power consumption is too great under certain conditions, cells must be redesigned and re-simulated, further increasing the simulation and time burden required for complete power analysis for the chip.

In this example, a variety of conditions need to be taken into account, with the following values for each:

- Temperature: −40, 27, 85 (in °C)
- Voltage: 2.5 V ± 10 %
- Power mode: 0, 1, 2
- Input state: 0, 1, 2
- Bias: 5uA ± 10 %
- Process corners: FF/FS/TT/SF/SS
- RC extraction corners: 1, 2, 3, 4, 5, 6, 7, 8, 9

The total number of combinations required to achieve full coverage of these conditions is 13,122. For a simulation time of one day on a cluster of 200 machines with 200 simulator licenses, it would take over two months to complete one full analysis of this design. It is easy to see how the addition of more variables quickly increases the number of combinations to well above 100,000.

To make the problem tractable, the number of combinations being simulated needs to be reduced. One way to do this is to use design expertise to determine combinations that are unlikely to have adverse power consumption. However, even with this approach, the number of combinations can still be very large. For the remaining combinations, the number of corners to be analyzed needs to be as large as possible.

To achieve this, Fast PVT is used to adaptively simulate and predict the power consumption under all of the required process/environment/state conditions. Fast PVT reduces the total number of simulations for covering the 13,122 combinations to 643, providing approximately a 20x simulation savings.

In summary, power analysis is key to designing successful mobile SoCs, but it is important that variation effects are analyzed across process, environmental, and power state conditions to ensure that the design stays within power constraints. The method chosen to do this must reconcile the tradeoff between thoroughness, available time, and computing resources.

2.7 Fast PVT: Discussion

We now examine Fast PVT in terms of the set of qualities required for a PVT technology outlined in the introduction.

1. **Rapid**: Fundamentally, Fast PVT can be fast because it learns about the mapping of PVT variables to output performances on-the-fly, and takes advantage of that knowledge when choosing simulations. As the benchmarks

demonstrated, Fast PVT verification has speedups averaging 4–5x; with speedup up to 22x.

2. *Reliable*: Fast PVT is reliable because it uses SPICE in the loop, nonlinear modeling technology, and takes measures to fill blind spots in the model. In benchmark results, Fast PVT finds the worst-case corners in 100 % of cases across a broad range of real-world problems.

In summary, Fast PVT is simultaneously rapid and reliable, which makes possible a highly efficient and practical PVT design flow.

2.8 Conclusion

In modern semiconductor processes, the need to properly bracket variation has caused the number of possible PVT corners to balloon. Rather than a handful of corners, designers must test against hundreds or even thousands of possible corners, making PVT-based design and verification exceedingly time-consuming.

This chapter described various possible flows to handle PVT variation, and various ways to find worst-case PVT corners. It then presented the Fast PVT approach, which casts PVT verification and corner extraction as a global optimization problem, then solves the problem using model-building optimization (MBO). Benchmark results verified that Fast PVT delivers good speedups while finding the true worst-case PVT corners in all benchmark cases.

Fast PVT enables a rapid PVT design flow, via fast extraction of worst-case PVT corners and fast verification. It reduces overall design time and improves reliability over conventional methods. This in turn promotes the reliable development of more competitive and more profitable products.

Appendix A: Details of Fast PVT Verification Algorithm

Detailed Algorithm Description

We now give a more detailed description of the Fast PVT algorithm. We do so in two parts: first, by showing how we recast the problem as a global optimization problem, then how this problem can be quickly and reliably approached with an advanced model-building optimization technique.

We can cast the aim of finding worst-case corners as a global optimization problem. Consider x as a point in PVT space, i.e. a PVT corner. Therefore, x has a value for the model set or for each device type if separate per-device models are used, V_{dd}, R_{load}, C_{load}, temperature T, etc. We are given the discrete set of N_C possible PVT corners $X_{all} = \{x_1, x_2, \ldots, x_{NC}\}$, and a SPICE-simulated output

Fig. 2.11 Fast PVT
verification algorithm

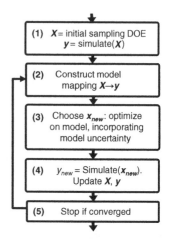

response to each corner $f(x)$. We aim to find x^*, the optimal PVT corner which gives the minimum or maximum $f(\mathbf{x})$, depending on the output. Collecting this together in an optimization formulation, we get:

$$\mathbf{x}^* = argmin\, (f(\mathbf{x}))$$
$$\text{subject to } x \text{ in } X_{all}$$

Now, given the aims of speed and reliability, the challenge is to solve the global optimization problem with as few evaluations of $f(x)$ as possible to minimize simulations, yet reliably find the x^* returning the global minimum, which is the true worst-case corner.

Fast PVT approaches this optimization problem with an advanced model-building optimization approach that explicitly leverages modeling error.

We now detail the steps in the approach, as shown in Fig. 2.11.

Step 1: Raw initial samples: Fast PVT generates a set of initial samples $X = X_{init}$ in PVT space using design of experiments (DOE) (Montgomery 2004). Specifically, the full set of PVT corners is bounded by a hypercube, then DOE selects a fraction of the corners of the hypercube in a structured fashion.

Simulate initial samples: Run SPICE on the initial samples to compute all initial output values: $y = y_{init} = f(X_{init})$.

Step 2: Construct model mapping $X \rightarrow y$: Here, Fast PVT constructs a regressor (an RSM) mapping the PVT input variables to the SPICE-simulated output values. The choice of regressor is crucial. Recall that a linear or quadratic model makes unreasonably strong assumptions about the nature of the mapping. We do not want to make any such assumptions—the model must be able to handle arbitrarily nonlinear mappings. Furthermore, the regressor must not only predict an output value for unseen input PVT points, it must be able to report its confidence in that prediction. Confidence should approach 100 % at points that have previously been simulated, and decrease as distance from simulated points increases.

An approach that fits these criteria is Gaussian process models (GPMs, a.k.a. kriging)(Cressie 1989). GPMs exploit the relative distances among training points and the distance from the input point to training points while predicting output values and the uncertainty of the predictions. For further details on GPMs, we refer the reader to Appendix B.

Step 3: Choose new sample x_{new}: Once the model is constructed, we use it to choose the next PVT corner x_{new} from the remaining candidate corners $X_{left} = X_{all} \backslash X$. One approach might be to simply choose the x that gives minimum predicted output value $g(x)$:

$$x_{new} = argmin(g(x)) \text{ subject to } x \text{ in } X_{left}$$

However, this is problematic. While such an approach optimizes $f(x)$ in regions near where worst-case values have already been simulated, there may be other regions with relatively fewer simulations, which have different simulated values than model predictions. These are model *blind spots*, and if such a region contained the true worst-case value, then this simple approach would fail.

GPMs, however, are aware of their blind spots because they can report their uncertainty. So, we can choose x_{new} by including uncertainty $s^2(x)$, where X_{left} is the set of remaining unsimulated corners from X:

$$x_{new} = argmin(h(g(x), \ s^2(x))) \text{s.t. } x \text{ in } X_{left}$$

where $h(x)$ is an infill criterion function that combines both $g(x)$ and $s^2(x)$ in some fashion. There are many options for $h(x)$, but a robust one uses least-constrained bounds (LCB) (Sasena 2002). This method returns the x_{new} that returns the minimum value for the lower-bound of the confidence interval. Mathematically, LCB is simply a weighted sum of $g(x)$ and $s^2(x)$.

Step 4: Simulate new sample; update: Run SPICE on the new sample: $y_{new} = f(x_{new})$. We update all the training data with the latest point: $X = X \cup x_{new}$, and $y = y \cup y_{new}$.

Step 5: Stop if converged: Here, Fast PVT stops once it is confident that it has found the true worst-case. Specifically, it stops when it has determined that there is very low probability of finding any output values that are worse than the ones it has seen.

Illustrative Example of Fast PVT Convergence

Figure 2.12 shows an example Fast PVT verification convergence curve, plotting output value versus sample number. The first 20 samples are initial samples X_{init} and y_{init}. After that, each subsequent sample x_{new} is chosen with adaptive modeling. The predicted lower bound shown is the minimum of all 95 %-confidence predicted lower bounds across all unsimulated PVT corners (X_{left}). The PVT

Fig. 2.12 Example of Fast
PVT convergence

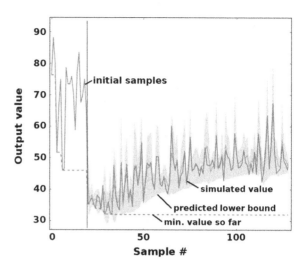

corner with this minimum value is chosen as the next sample x_{new}. That new
sample is simulated.

The dashed line in Fig. 2.12 is the minimum simulated value so far. We see that
immediately after the initial samples, the first x_{new} finds a significantly lower
simulated output value $f(x_{new})$. Over the course of the next several samples, Fast
PVT finds even lower simulated values. Then, the minimum value curve flattens,
and does not decrease further. Simultaneously, from sample number 20–40, we see
that the predicted lower bound hovers around an output value of 30, but then after
that, the lower bound increases, creating an ever-larger gap from the minimum
simulated value. This gap grows because X_{left} has run out of corners that are close
to worst-case, hence the remaining next-best corners are much higher than the
already-simulated worst-case. As this gap grows, confidence that the worst-case is
found increases further, and at some point we have enough confidence to stop.

Appendix B: Gaussian Process Models

Introduction

Most regression approaches take the functional form:

$$g(x) = \sum_{i}^{NB} w_i g_i(x) + \varepsilon$$

Where $g(x)$ is an approximation of the true function $f(x)$. There are N_B basis
functions; each basis function $g_i(x)$ has weight w_i. Error is ε. Because $g_i(x)$ can be
an arbitrary nonlinear function, this model formulation covers linear models,

polynomials, splines, neural networks, support vector machines, and more. The overall class of models is called generalized linear model (GLM). Model fitting reduces to finding the w_i and $g_i(x)$ that optimize criteria such as minimizing mean-squared error on the training data, and possibly regularization terms. These models assume that error ε is normally distributed, with mean of zero, and with no error correlation between training points.

In this formulation, the error distribution remains constant throughout input variable space; it does not reduce to zero as one approaches the points that have already been simulated. This does not make sense for SPICE-simulated data: the model should have 100 % confidence (zero error) at previously simulated points, and error should increase as one draws away from the simulated points. Restating this, the model confidence should change depending on the input point.

Towards Gaussian Process Models (GPMs)

We can create a functional form where the model confidence *depends* on the input point:

$$g(x) = \sum_{i}^{NB} w_i g_i(x) + \varepsilon(x)$$

Note how the error ε is now a function of the input point x. Now the question is how to choose w_i, $g_i(x)$, and $\varepsilon(x)$ given our training data X and y. A regressor approach that fits our criteria of using $\varepsilon(x)$ and handling arbitrary nonlinear mappings, is the Gaussian process model approach (GPMs, a.k.a. kriging). GPMs originated in the geostatistics literature (Cressie 1989) but have recently become more popular in the global optimization literature (Jones et al. 1998) and later in machine learning literature (Rasmussen and Williams 2006). GPMs have such a powerful approach to modeling $\varepsilon(x)$ that they can replace the first term of $g(x)$ with a constant μ, giving the form:

$$g(x) = \mu + \varepsilon(x)$$

In GPMs, $\varepsilon(x)$ is normally-distributed with mean zero, and variance represented with a special matrix R. R is a function of the N training input points X, where correlation for input points x_i and x_j is $R_{ij} = corr(x_i, x_j) = exp(-d(x_i, x_j))$, and d is a weighted distance measure $d(x_i, x_j) = \sum_{h=1n} \theta_h |x_{i,h} - x_{j,h}|^{p_h}$. This makes intuitive sense: as two points x_i and x_j get closer together, their distance d goes to zero, and therefore their correlation R_{ij} goes to one. Distance measure d is parameterized by n-dimensional vectors θ and p, which characterize the relative importance and smoothness of input variables. θ and p are learned via maximum-likelihood estimation (MLE) on the training data.

From the general form $g(x) = \mu + \varepsilon(x)$ which characterizes the distribution, GLMs predict values for unseen x via the following relationship:

$$g(x) = \mu + r^{\mathrm{T}} R^{-1}(y - 1\mu)$$

where the second term adjusts the prediction away from the mean based on how the input point x correlates with the N training input points X. Specifically, $r = r(x) = \{corr(x, x_1), \ldots, corr(x, x_N)\}$. Once again, this formula follows intuition: as x gets closer to a training point x_i, the influence of that training point x_i, and its corresponding output value y_i, will become progressively greater.

Recall we want a regressor to not just predict an output value $g(x)$, but also to report the uncertainty in its predicted output value. In GLMs, this is an estimate of variance s^2:

$$s^2(x) = \sigma^{2*} \left(1 - r^T R^{-1} r + \left(1 - 1^T R^{-1} r \right)^2 / \left(1^T R^{-1} 1 \right) \right)$$

In the above formulae, μ and σ^2 are estimated via analytical equations that depend on X and y. For further details, we refer the reader to (Jones et al. 1998).

GPM Construction Time

With GPMs, construction time increases linearly with the number of parameters, and as a square of the number of training samples. In practical terms, this is not an issue for 5 or 10 input PVT variables with up to ≈ 500 corners sampled so far, or for 20 input PVT variables and ≈ 150 samples; but it does start to become noticeable if the number of input variables or number of samples increases much beyond that.

In order for model construction not to become a bottleneck, the Fast PVT algorithm behaves as follows:

- Once 180 simulations have been reached, it only builds models every 5 simulations, rather than after every new simulation. The interval between model builds increases with the number of simulations (=max(5, 0.04 * number of simulations)).
- If Fast PVT has not converged by 1000 simulations, it simply simulates the rest of the full-factorial corners.

References

Cadence Design Systems Inc. (2012) Cadence® Virtuoso® Spectre® Circuit Simulator, http://www.cadence.com

Cressie N (1989) Geostatistics. Am Statistician 43:192–202

Huyer W, Neumaier A (1999) Global optimization by multilevel coordinate search. J Global Optim 14:331–355

Jones DR, Schonlau M, Welch WJ (1998) Efficient global optimization of expensive black-box functions. J Global Optim 13:455–592

Land AH, Doig AG (1960) An automatic method of solving discrete programming problems. Econometrica 28(3):497–520

Montgomery DC (2004) Design and analysis of experiments, 6th Edition, Wiley, Hoboken

Rasmussen CE, Williams CKI (2006) Gaussian processes for machine learning, MIT Press, Cambridge, MA

Sasena MJ (2002) Flexibility and efficiency enhancements for constrained global optimization with kriging approximations, PhD thesis, University of Michigan

Solido Design Automation Inc. (2012) Variation Designer, http://www.solidodesign.com

Chapter 3
A Pictorial Primer on Probabilities

Intuition on PDFs and Circuits

Abstract This chapter aims to build intuition about probability density functions (PDFs), Monte Carlo sampling, and yield estimation. It has an emphasis on graphical analysis as opposed to equations. Such intuition will help in many design scenarios, when one is observing actual PDF data in the form of scatterplots, histograms, and normal quantile (NQ) plots.

3.1 Introduction

This chapter starts with very basic descriptions of probability distributions in one dimension, and then moves to multiple dimensions. It then describes statistical variation in circuits and Monte Carlo sampling. It builds insight on non-Gaussian distributions, the propagation of distributions through linear and nonlinear functions, and how this insight can be used to understand circuit behavior. The remaining sections describe histograms and density estimation, statistical estimators like average and yield, and NQ plots.

3.2 One-Dimensional Probability Distributions

A probability distribution (or just distribution) is simply a function that describes how probable different input values are. As we will see, this simple description is deceptively powerful.

Consider the statement: "$x = 10$, with 80 % probability, otherwise $x = 20$", as shown in Fig. 3.1a. This is a probability distribution: it describes how probable different input values (x) are. In this case, x can take just two different values, and

T. McConaghy et al., *Variation-Aware Design of Custom Integrated Circuits:*
A Hands-on Field Guide, DOI: 10.1007/978-1-4614-2269-3_3,
© Springer Science+Business Media New York 2013

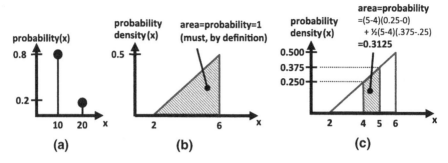

Fig. 3.1 **a** A discrete probability distribution **b** a continuous probability distribution **c** computing a probability in a continuous probability distribution

all other values have probability zero. Because it has a finite set of possible input values, it is a *discrete* probability distribution. By definition, in a discrete probability distribution, the sum of all probabilities is 1.0.

Figure 3.1b is an example of a *continuous* probability distribution. The input variable *x* can take continuous values. Note how the *y*-axis is not probability, but rather probability *density*. What this means is that to get a probability, one must compute an *area* (integrate) under the probability density curve. By definition, the area under the whole curve must have a probability of 1.0. If we compute the area for a *range* of *x*-values, that is the probability that the given range of *x*-values will occur. For example, Fig. 3.1c computes that a value of *x* between 4.0 and 5.0 will occur with probability 31.25 %.

It is worth emphasizing that for continuous probability distributions, the idea of a *probability* for a given input *x* value is meaningless. For example, if we had an *x*-value of 5.0, then since it's just one value then we take its range as 5.0–5.0, having a width of $5.0 - 5.0 = 0.0$. With a width of 0.0, then the area (and probability) is also 0.0. For continuous-valued distributions, it is the *probability density* that is meaningful for a given value of *x*, not *probability*. It works the other way too: "density" is meaningless in discrete-valued distributions.

The label "PDF" is shorthand for "probability distribution function" (for discrete or continuous distributions), or for "probability density function" (for continuous distributions only).

Figure 3.2 shows two famous and widely-used continuous PDFs. Figure 3.2a is a Gaussian, or normal, PDF. Normal PDFs have two parameters: the mean (μ) characterizes what values *x* tends to be near to, and the standard deviation (σ) characterizes the degree of dispersion from the mean. 68.27 % of the *x* values are within one standard deviation of the mean, 95.45 % of values lie within two standard deviations, and 99.73 % of values lie within 3 standard deviations. The actual formula is $pdf(x) = 1/(\sigma^*\sqrt{2\pi})exp\left(-(x-\mu)^2/(2\sigma^2)\right)$. Shorthand notation for a normal distribution is N(μ, σ). Figure 3.2a shows the normal distribution in N(0,1) which is *standard form*, when $\mu = 0$, and $\sigma = 1$.

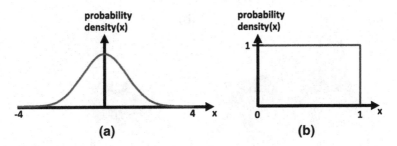

Fig. 3.2 a Gaussian (normal) probability density function, and **b** Uniform probability density function

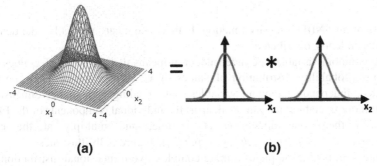

Fig. 3.3 a Two-dimensional Gaussian PDF with mean = (0,0), standard deviation = (1,1) and no correlation. Since its variables are independent, its *pdf(x₁, x₂)* is equivalent to **b** *pdf(x₁)* * *pdf(x₂)*

Figure 3.2b is a uniform PDF. Uniform PDFs have two parameters: the minimum and maximum value. Outside the range of minimum and maximum values, the density is 0.0, and within the range, the density is a constant value such that overall area (probability) is 1.0. Figure 3.2b shows the uniform PDF with min = 0 and max = 1. In general, a uniform PDF follows the form:

$$pdf_{uniform}(x) = \left\{ \begin{array}{ll} \left(\dfrac{1.0}{max - min}\right) & if \quad x \epsilon [min, max] \\ 0.0 & otherwise \end{array} \right\}.$$

3.3 Higher-Dimensional Distributions

Distributions may have more than one input dimension. Figure 3.3a is an example of a two-dimensional continuous PDF: x_1 and x_2 are the input random variables, and the z-axis is the probability density. The density function of Fig. 3.3a is an

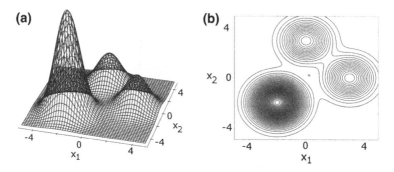

Fig. 3.4 Mixture-of-Gaussians two-dimensional PDF **a** surface-plot version **b** contour-plot version

example of an "NIID" density function: both x_1 and x_2 are **n**ormally, **i**dentically, and **i**ndependently **d**istributed[1].

When random variables are independent, it means that the value of x_1 does not affect the probability distribution of values of x_2, or vice versa. They are not correlated. Independent random variables make dealing with higher dimensionality far more tractable: one can break apart the individual components of the PDF, deal with them one dimension at a time, and multiply at the end: $pdf(x_1, x_2, \ldots, x_n) = pdf(x_1)^*pdf(x_2)^{*\cdots*}pdf(x_n)$. Figure 3.3b illustrates.

Figure 3.4a is an example of a more complex two-dimensional distribution, in this case a *mixture* of three Gaussians. It is a *multimodal* distribution because it has more than one peak. This means that when drawing random points from it, it will tend towards one of the three peaks: the most likely peak is at $(x_1, x_2) = (-2, -2)$, with other peaks being (0, 3) and (3, 0). The distribution of Fig. 3.4(a) is not NIID for several reasons. It is not normally distributed because it does not follow the 2-dimensional form of a normal distribution. Its random variables x_1 and x_2 are not identically distributed because they follow different distributions. Finally, x_1 and x_2 are not independent: the value of x_1 affects the probability of different values that x_2 might take; and we cannot compute the PDF of x_1 and x_2 independently for an overall density value.

Figure 3.4b shows the same distribution as Fig. 3.4a, just in contour plot form. For illustrating certain characteristics of PDFs, contour plots are often better than surface plots. For example, the spatial distribution across x_1 and x_2 is easier to read in the contour plot in Fig. 3.4b, compared to the surface plot of Fig. 3.4a.

We can compute probabilities in two or more dimensions as well: it is integration under the PDF curve. For example, the probability that $(x_1 < 0.0$ and $x_2 < 0.0)$ in

---

[1] For further clarification: Two identically distributed distributions have exactly the same distribution shape (e.g., normal) and distribution parameters (e.g., same mean and standard deviation). Two "independently distributed" distributions have no correlation with each other; if we have a value for one distribution, that value does not change what the probable values for the other distribution might be.

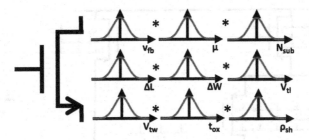

Fig. 3.5 There are ≈ 10 independent random variables to model variation of a device in the widely-used BPV model (Drennan and McAndrew 2003)

Fig. 3.3a is the volume under the curve, which works out to 0.25. This is a simple example with an analytical solution. But consider the challenge of calculating integrals on Fig. 3.4a, but with more complex boundaries and with 1,000 dimensions rather than 2. As we see later, this starts to approximate the central challenge for yield estimation problems (and begins to hint where tools may help).

3.4 Process Variation Distributions

In modern PDKs supplied by the foundries, each locally–varying process parameter of each device is modeled as a distribution. On top of this, global process variation is either modeled as a distribution, or as "corners" bounding a distribution.

We aim to give some intuition about modeling process variation by describing one statistical model of variation, as an example. Of course, other models exist, and this book does not preclude their use. What really matters from a design perspective is simply "what's in the PDK". With that in mind, we will describe a model that is widely used, accurate, and logically designed: the back-propagation of variance (BPV) model (Drennan and McAndrew 2003). The model continues to be refined, for example (McAndrew et al. 2010) and (Li et al. 2010).

At the core of the BPV model is one n-dimensional NIID (normally, identically, independently-distributed) distribution for each device, and an n-dimensional NIID distribution for global variation. Each of the n variables captures how underlying, physically independent process parameters change. These parameters are flatband voltage V_{fb}, mobility μ, substrate dopant concentration N_{sub}, length offset ΔL, width offset ΔW, short channel effect V_{tl}, narrow width effect V_{tw}, gate oxide thickness t_{ox}, and source/drain sheet resistance ρ_{sh}. This is the list from (Drennan and McAndrew 2003) but more physical parameters may be added to the mix, and this often happens in practice. Figure 3.5 illustrates.

Variation of these physical process parameters leads directly to variation (in silicon, and in simulation) of a device's electrical characteristics like drain current I_d, input voltage V_{gs}, transconductance g_m, and output conductance g_d.

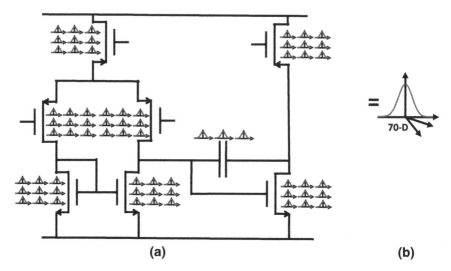

Fig. 3.6 **a** With ≈ 10 random variables for each transistor, even a simple Miller OTA has $7*10 = 70$ random variables for just local variation; and several more for global variation and the capacitor variation. **b** A simplified way of thinking about the 70-dimensional distribution

Some models of process variation are not normally distributed; for example, some random variables are lognormally or uniformly distributed. In other models of variation, the random variables are not independent (e.g., they have correlations). In both cases, one can apply mathematical transformations to the distribution such that the distribution becomes NIID (and therefore easier to work with).

As a rule-of-thumb, on modern foundries' PDKs, one can expect at least 10 process variables per device (and perhaps 10 global process variables). Put another way, *the number of variables is 10 times the number of devices*. This has wide ramifications for variation-aware design, because it affects which algorithms and tools may be useful. Even a 7-transistor Miller operational transconductance amplifier (OTA) has 70 local process variables, as Fig. 3.6a shows. A 100-device circuit like a big opamp has 1000 variables, and a 10,000 device circuit like a phase-locked loop (PLL) has 100,000 variables. We will see the effect of this number throughout this book.

Figure 3.6b shows a technique we will use to illustrate higher-dimensional distributions.

3.5   From Process Variation to Performance Variation

One distribution can *map* to another via *functions*. In circuits, the distributions of all physical local variations, and global variations, map directly to variations in a circuit's performance characteristics (e.g., power, delay).

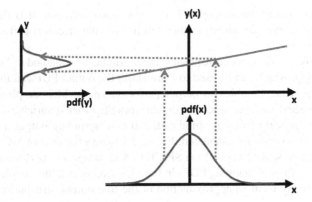

Fig. 3.7 This is a simple example showing how a PDF (e.g. process variation) is transformed into another PDF (e.g., performance variation), via an intermediate function. In this case, a Gaussian-distributed variable *x* (*bottom*) has a linear mapping *y(x)* to *y*, and therefore *y* is Gaussian distributed too, but is shifted and scaled (*left*)

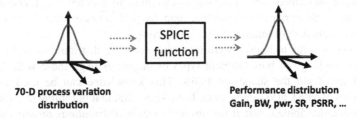

Fig. 3.8 The high-dimensional PDF of process variations maps into the high-dimensional PDF of performance variation, via simulation (or silicon)

Figure 3.7 is a simple one-dimensional example showing how an input PDF is mapped to a new PDF. In this example, the final PDF is just a shifted and scaled version of the original PDF, but more nonlinear transformations may happen. (Sect. 3.7 will elaborate on different types of transformations.)

Figure 3.8 generalizes this concept to a more general circuit setting, where the input is a high-dimensional PDF describing process variations, the mapping function is a SPICE-like circuit simulator, and the output PDF describes performance variations. The output PDF has one dimension for each output performance (gain, BW, etc.).

3.6 Monte Carlo Sampling

Thanks to the statistical models that are part of modern foundry-supplied process design kits (PDKs), we have direct access to the distributions of the process variables. However, we do not have direct access to the distributions of the output performances, mainly because SPICE is a black-box function. It is these output distributions that we care about, as they tell us how the circuit is performing, and what the yield is.

Fortunately, we have at our disposal is the tried-and-true method of Monte Carlo (MC) sampling, which can be used to learn about the output PDF and the mapping from input process variables to output performance. In MC sampling, samples (points) in process variable space are drawn from the process variable distribution. Each sample (process point) is simulated, and corresponding output performance values for each sample are calculated. Figure 3.9 shows the flow of MC samples in process variation space, mapped via SPICE to MC samples in performance space. This is like the flow of mapping PDFs in Fig. 3.8, except that the samples in output space collectively form an approximation of the true output distribution[2].

3.7 Interpreting Performance Distributions

As we have described, one distribution can map to another via functions. In circuits, physical parameter variations map to performance variations, via the "blackbox" function of simulation (or of silicon).

This section aims to build insight into what *mapping* distributions is about, by building knowledge of how different types of mappings (e.g., those with discontinuities) transform the shapes of PDFs. This knowledge can be used to gain insight from Monte Carlo sample data. If we know that that our initial distributions are Gaussian-distributed, and *if we observe certain distributions on our outputs, then we can make inferences about the nature of the mapping.* This is highly useful: it provides a way to get insight about the mapping and the nature of the circuit's behavior, despite the blackbox nature of SPICE.

In all the figures, the reader may treat the input variable x as a process parameter (e.g., flatband voltage), and the output variable y as a circuit performance parameter (e.g., power) or a device performance parameter (e.g., transconductance).

Figure 3.7 illustrates a Gaussian PDF, passed through a linear mapping. The result is a Gaussian PDF as well. The final PDF, however, is shifted (has a different mean) because the linear function does not cross at the origin (0,0); and it is scaled (has a different standard deviation) because the slope of the linear function is not 1.0. If we observe an output's distribution to be approximately Gaussian-distributed, then we can assume that the mapping is approximately linear (at least in the

---

[2] In the statistics literature, "Monte Carlo methods" are a broad set of algorithms, where the unifying element is that each algorithm has some randomization. While we use the label "sample" to mean a single point in process variable space drawn from a distribution or its corresponding output performance values, in the statistics literature such a point is an "observation" and a "sample" is a *set* of observations. The circuits community tends to use "sample" in the way we do.

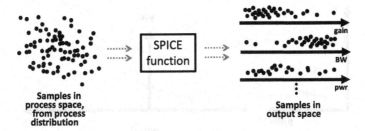

Fig. 3.9 The flow of Monte Carlo sampling, from process variation space via SPICE to output space

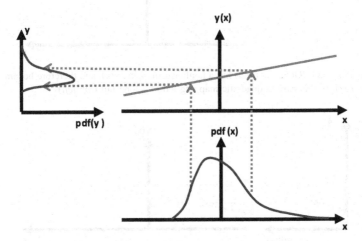

Fig. 3.10 *Non*-Gaussian PDF(x) (bottom) has a linear mapping $y(x)$ to y, and therefore retains its shape, but is shifted and scaled (*left*)

range of variation tested). Put another way, changes in output performance are proportional to changes in process variation parameters. Linear or near-linear mappings show up (unsurprisingly) in circuits with just linear devices, such as passive filters. They may also show up in some outputs of some circuits where the intent is linear behavior (e.g., opamps); though only when the process variation is small enough so that the nonlinear transforms do not kick in (e.g., devices do not turn off due to massive variation).

Figure 3.10 illustrates the mapping from a *non*-Gaussian PDF, through a linear mapping, and the result is a PDF with the same shape as the original PDF, just shifted and scaled. The point of this drawing is that linear mappings affect both Gaussian distributions and non-Gaussian distributions in the same fashion—shifting and scaling, but retaining the original shape.

Figure 3.11 illustrates how a Gaussian PDF becomes transformed into a *non*-Gaussian PDF due to a nonlinear mapping. In this case, the nonlinear mapping is a quadratic, which leads to a long tail. Therefore, if we observe that an output has a longer tail, we can know that the mapping is quadratic; i.e. there is a

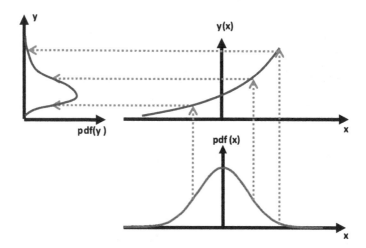

Fig. 3.11 Gaussian PDF(*x*) (bottom) has a nonlinear mapping *y(x)*, and therefore becomes non-Gaussian (*left*). In this case, a quadratic mapping led to a long tail (*top left*)

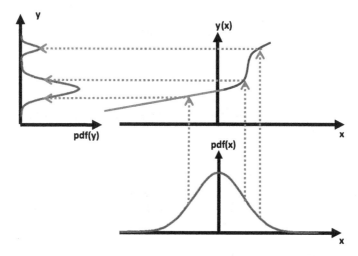

Fig. 3.12 Gaussian PDF(*x*) (bottom) has a nonlinear mapping *y(x)*, and therefore becomes non-Gaussian (*left*). In this case, a discontinuous mapping led to a bimodal distribution (*top left*)

disproportionate effect on the output as some process parameters get larger or smaller. As an example, read current of a bitcell typically has a quadratic response across most values of process variables.

Figure 3.12 illustrates how a Gaussian PDF becomes transformed to a *non-Gaussian* PDF as well, due to a nonlinear mapping. In this case, the nonlinear mapping has a sharp change, nearly a discontinuity, then resumes a more smooth mapping. This near-discontinuity causes a region of the output *y*-space to be skipped, and the final result is a bi-modal distribution. Therefore, if we observe

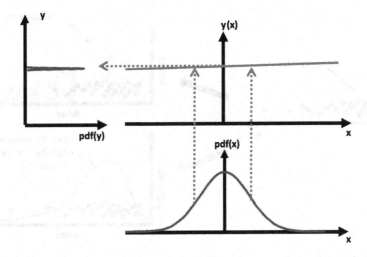

Fig. 3.13 From a Gaussian PDF(x) (*bottom*), the mapping has basically no response to *x*, and therefore the output distribution is basically constant (*left*)

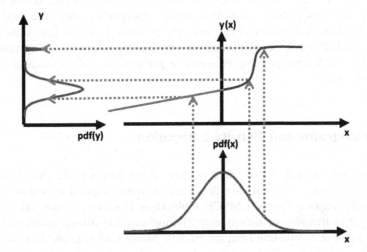

Fig. 3.14 Here, the mapping is a combination of a near-discontinuity, followed by a flat response in the region after the discontinuity. The resulting distribution has a second mode (*top left*) that is near-constant

that an output distribution is bimodal, then we know that the mapping has a near-discontinuity. If we observe more than two modes (two "humps") in the output PDF, then we know that there are even more discontinuities.

Figure 3.13 is an example where the mapping from process variable *x* to output *y* is nearly flat. This results in an output distribution that is nearly constant (a near-"spike").

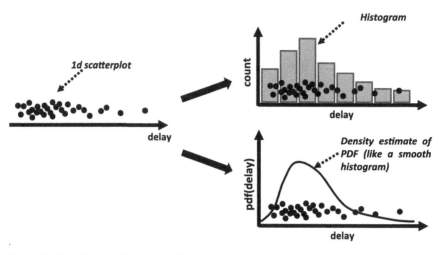

Fig. 3.15 Visualizing 1-dimensional distributions

Figure 3.14 is an example where the mapping combines a near-discontinuity and a flat response, which results in a bimodal distribution where one mode is a constant value. A common circuit example is when variation gets sufficiently extreme such that a device gets turned off, it essentially destroys the operation of the circuit, with corresponding drop-offs in performance values (such as power going to zero).

3.8 Histograms and Density Estimation

We previously described that while we have direct access to the distribution of process variation, we do not have direct access to the output distribution, in part because the mapping function, SPICE, is blackbox. However, we can still *estimate* the output distribution in various ways. Histograms and density estimation estimate the output distribution from the outputs' Monte Carlo sample data. We now elaborate.

There are many ways to visualize a probability distribution based on Monte Carlo samples. Figure 3.15 illustrates typical visual representations of a 1-dimensional distribution: as a scatterplot of the samples, or in some fashion that aggregates the samples, such as binning them into groups and counting the number of samples per bin—a histogram.

A histogram can be seen as a *density estimation* technique: from the existing data, the binning process constructs a discrete distribution, as shown in Fig. 3.15 top right.

We can make a "smooth" histogram by *continuous-valued density-estimation*. Let us explain by a simple example: assume a Gaussian distribution, and characterize it via estimates of mean and standard deviation from the samples. This is a density estimate of a (continuous) probability density function, or PDF. A PDF returns a probability *density* value for any input value. To calculate a *probability*, one measures the area under the PDF across a range of input values (i.e. integrates); Sect. 3.9 elaborates on this for yield estimation.

Of course, a Gaussian model of a distribution does not do a good job capturing characteristics of many distributions commonly encountered in custom ICs, such as long tails (as shown), or multi-modal distributions where there is more than one "hump". Fortunately, the field of density estimation is long established and there are many standard techniques to make better estimates.

Given a set of samples, a good density estimate maximizes the probability of the data, given the density model. One very common approach is to use a mixture of two or more Gaussian distributions, to make a so-called "Gaussian Mixture Model". In the example of two Gaussians, one would have to estimate the mean and standard deviation for each Gaussian, as well as the relative weight of one distribution over the other.

One can take this to a logical extreme and have a Gaussian centered at each datapoint, with a well-chosen standard deviation for each Gaussian; this technique is called "Kernel Density Estimation" (KDE) (Parzen 1962; Rosenblatt 1956). There are plenty of other techniques too; "density estimation" remains an active research field (Hastie et al. 2009).

3.9 Statistical Estimates of Mean, Standard Deviation, and Yield

An "estimate" is exactly what one would expect: a guess of what a value is, given some (but not "all") data. A "statistical estimate" has a more precise meaning: it is an estimate of a property of a distribution, given samples from the distribution. Equivalently, the "population value" is the true value of the property, and the "sample value" is the value of the property estimated from the sample data.

Mean and standard deviation are the most-discussed properties of distributions. Monte Carlo (MC) sampling is a common way to estimate their values.

Given a set of MC samples $\{x_1, x_2, \ldots, x_i, \ldots, x_N\}$, here are the estimators for mean and standard deviation:

- *Average, $\bar{x}$,* is the estimate of the *mean, μ.* Average is computed as $\bar{x} = \dfrac{1}{N}\sum_{i=1}^{N} x_i$

- *Sample standard deviation, s,* is the estimate of the (*population*) *standard deviation, σ;* where $s = \sqrt{\dfrac{1}{N-1}\sum_{i=1}^{N}(x_i - \bar{x})^2}$.

Fig. 3.16 Computing yield
given a specification. Going
from spec to yield is
integration. In the 1-D case,
integration is equivalent to
computing the area under the
PDF in the range where the
output is feasible

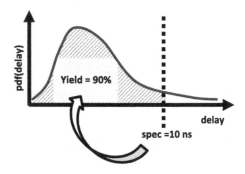

It turns out that we can estimate the distribution properties in different ways,
with different benefits for each approach. For example, another estimate for
population standard deviation, σ, is the *standard deviation of the sample*[3], s_N,
computed as $s_N = \sqrt{\dfrac{1}{N}\sum_{i=1}^{N}(x_i - \bar{x})^2}$. Whereas s_N tends to estimate σ too low for a
small number of MC samples, s tends to estimate σ too low for a larger number of
MC samples. The most common estimator is s.

Yield is another property of distributions. It has very specific physical meaning
for circuit designs. Physically, yield is the percentage of chips that meet specifi-
cation(s). Figure 3.16 illustrates yield in terms of distributions. Yield is the area
under the PDF in the region where the pdf meets the target specification(s), i.e.
where it is *feasible*. In the example of Fig. 3.16, there is a single specification:
delay must be ≤ 10 ns. The area under the PDF where delay ≤ 10 ns is 0.9, or a
yield of 90 %.

Yield can also be estimated in different ways. Here are a couple of common
ways:

- *Monte Carlo sampling.* Given a set of MC samples $\{x_1, x_2, \ldots, x_i, \ldots, x_N\}$ and
  an indicator function $I(x)$ which outputs a 1 if a sample x meets spec(s) and 0
  otherwise. Then, $yield = \frac{1}{N}\sum_{i=1}^{N} I(x_i)$.

- *Density estimation.* This involves making a density estimate of the PDF, then
  yield is the volume under the PDF where it meets specifications. For some
  PDFs, this may be analytical (e.g., using the Gaussian CDF[4] formula); whereas
  for other PDFs this may require a numerical technique (e.g., drawing a large
  number of MC samples from the PDF, or applying bisection search).

Yield itself may have different units. In the estimates above, yield is in a range
0.0–1.0. Yield may be represented in percentage form, by multiplying the

---

[3] We agree, the terminology is somewhat confusing!

[4] CDF = Cumulative Distribution Function. *CDF(x)* is the area under the PDF from $-\infty$ to x.

Fig. 3.17 The percentile function computes a specification, given a yield target

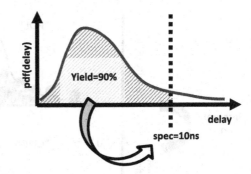

fractional estimates above by 100 %. As described in the Introduction chapter, yield may also be in units of probability of failure, or sigma.

It turns out that the *percentile* function is the inverse function to the yield function, if there is one output specification. Given a PDF, the yield function takes in a specification and outputs a yield estimate; and the percentile function takes in a yield target and outputs a specification. Figure 3.17 illustrates. This has broad usage, for example computing the "3-sigma value" of a given output; i.e. the value of an output if it had 99.86 % yield.

In circuit design, yield may be measured at different points of the design and manufacturing flow. *Parametric yield* refers to the yield estimate using circuit simulation. *Manufacturing yield* is based on the actual number of manufactured dies that meet specifications. Ideally, these measures would give identical values, but that occurs only if simulation-level modeling sufficiently captures all the electrical and physical effects. This requires good models of global variation (wafer-to-wafer and die-to-die), local variation (within-die), temperature and other environmental effects, layout parasitics and other layout effects, and so on. This is the challenge of *silicon calibration*. In practice, this is very difficult to achieve perfectly; however, designers do find that improvements in parametric yield usually lead to improvements in manufacturing yield.

Partial yield is the yield of a single output performance measure. *Overall yield* is the yield of all performance measures at once, i.e. the percentage of circuits that meet all specs. The relation between overall yield and partial yield depends on how closely different output values correlate. On one hand, if there is a perfect correlation between all output measures (e.g., when an MC sample fails on one output, it also fails on the other, and vice versa), then the partial yield is equal to overall yield. At the opposite extreme, if there is no correlation between any output measures, then the overall yield is much lower: $yield_{overall} = yield_1 * yield_2$. For example, if $yield_1 = 0.90$ and $yield_2 = 0.80$, then $yield_{overall} = 0.72$. In general, since overall yield is what really matters to the design, and since it is very difficult to know the relation among outputs in advance, we advise a focus on overall yield.

It turns out to be challenging to estimate *overall* yield using density estimation, when there are >1 outputs. Appendix A of Chap. 4 discusses the challenge, and describes a resolution.

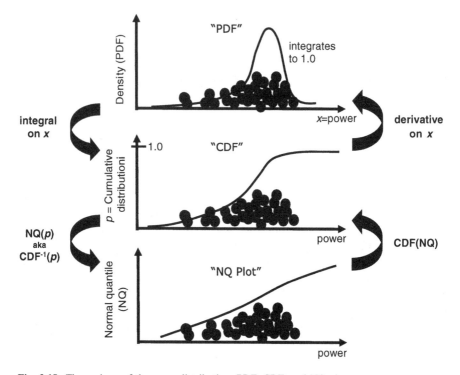

Fig. 3.18 Three views of the same distribution: PDF, CDF, and NQ views

3.10 Normal Quantile Plots

A Normal Quantile (NQ) plot is an alternative view of a distribution that facilitates easy comparison to Gaussian distributions, and enables detailed analysis of the distribution's tails, which is very useful in high-sigma analysis.

Figure 3.18 shows three views of the same distribution. The only difference between each view is the y-axis. Each view has its unique advantages and disadvantages. Crucially, we can transform among the views in a fairly straightforward fashion. We now examine each view, one at a time.

PDF: The top plot in Fig. 3.18 has the PDF (probability density function) on the y-axis. This is the plot used in the examples earlier in this chapter. The area under the PDF must integrate to 1.0. The area under the curve between a given range of x-values is the probability of that range of x-values. The PDF curve is intuitive because its y-value for a range of x values is proportional to the number of samples seen in that range of x-values. "Discretized" PDFs—histograms—are intuitive and easy to compute.

But the PDF view has disadvantages. First, lower-probability regions run extremely close to the x-axis, and have such small values that it is difficult to distinguish one low-probability value from another (e.g. 1e-4 from 1e-5), because they are both hugging the x-axis. This is troublesome for high-sigma analysis. Second, the "best-behaved" or "typical" distributions, namely Gaussian distributions, have a highly nonlinear bell-shaped curve. This means that small nonlinear distortions of that curve are difficult to identify, and in general the nonlinear curve is hard to work with. Compare this to many subdomains of electrical engineering, from circuit analysis to control theory, where the "best-behaved" models or systems are linear or linearized models; and linear techniques are key analysis and design tools. Nonlinear bell-shaped curves are not amenable to such analysis.

CDF: The middle plot in Fig. 3.18 has the CDF (cumulative distribution function) on the y-axis. A CDF value at x is the area under the PDF from ∞ to x. That is, $CDF(x) = \int_{-\infty}^{x} PDF(x)dx$. Intuitively, the y-axis can be viewed as the probability p that up to a given x-value can occur. CDFs are highly useful for computing yield: CDF values are equal to yield values for "$\leq$" specs; and equal to 1-yield for "$\geq$" specs. Accordingly, one may inspect CDF plots to see the tradeoff between yield values and spec values. To go backwards from CDF to PDF, one takes the derivative.

However, CDFs have the same disadvantages of PDFs: it is hard to distinguish among low-probability regions, and Gaussian distributions have highly nonlinear curves which impedes analysis.

NQ: Suppose there were a view of a distribution that overcomes some of the disadvantages described, while keeping key benefits. Specifically, suppose this view had the following properties:

- The "best-behaved" distributions (Gaussian) are linear curves, and the larger the deviation from Gaussian the more nonlinear the curve. Different types of deviations indicate different nonlinearities.
- One can directly see the tradeoff between yield and performance.
- One can easily distinguish among different low-probability values.

Remarkably, such a view of a distribution exists: it is an *NQ plot*. An example NQ plot is shown in the bottom of Fig. 3.18. An NQ plot is like a CDF, but the y-axis is warped "just right" such that if the underlying distribution is Gaussian, then the NQ appears as linear. The specific function doing the warping is the "normal quantile" function, also known as the "inverse CDF of the Gaussian", or "probit" function. It takes in a CDF value (a probability p), and outputs an NQ value. The NQ function is not available in closed form, so must be numerically computed as $NQ(p) = \sqrt{2}erf^{-1}(2p - 1)$, where erf^{-1} is the "inverse error function" and is typically provided in software packages and libraries.

The "normal quantile" value has a specific interpretation that aids intuition: the number of standard deviations away from the mean, if the distribution was Gaussian. Sigma is a unit for yield that is often simpler to use than percent yield or probability of failure; and one quickly can build intuition about the units. Specifically, to build intuition:

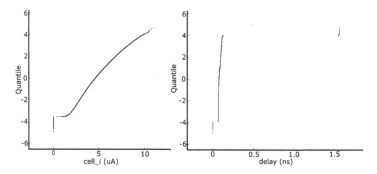

Fig. 3.19 NQ plot for bitcell read current (*left*) and sense amp delay (*right*)

- 2 sigma represents about 1 failure in 20
- 3 sigma is about 1 in 1000
- 4 sigma is about 1 in 50 thousand
- 5 sigma is about 1 in a million
- 6 sigma is about 1 in a billion

Certain circuit types typically have particular target ranges of sigma values. For example, non-replicated circuits like analog circuits are typically targeted for meeting 2–4 sigma, and bitcells are typically targeted for much higher yields around 6 sigma.

Density estimation: Recall the earlier discussion on density estimation, referring to estimating the probability density function (PDF) from data. There is another way to do density estimation: we can curve-fit the mapping from output performance value to NQ value, or vice versa. This may be considerably easier, because the NQ curves will typically be far less nonlinear than the nonlinear curves found in PDF space (even for Gaussian functions). When curve-fitting NQ data, to guarantee mathematical consistency, the mapping from output value to NQ value must be monotonically increasing. (McConaghy 2010) explores curve-fitting on NQ space in more detail, and Chap. 5 employs a piecewise-linear (PWL) density curve for statistical system-level analysis.

Random variates: A model mapping NQ values to output values has an additional benefit: it can be used to generate random values for the output value according to whatever arbitrary distribution the model has. That is, it can be used as a "random variate generator". Here's how: (1) draw a random value from a uniform distribution in the range [0,1]; (2) convert it to NQ via CDF^{-1}; and finally (3) run it through the model to get the output value.

NQ for insight: We can use the teachings of this chapter to show how inspection of NQ plots gives excellent insight into circuits. Figure 3.19 shows NQ plots from real circuit simulation data. The chapter on high-sigma analysis (Chap. 5) has the details of the setup; here we focus on the analysis. Each plot has approximately 1 million MC samples simulated. Each dot is a different MC sample.

Let us first look at Fig. 3.19 left, which shows an NQ plot for bitcell read current. First, we note that the curve does not follow a straight line, meaning it is not Gaussian-distributed. We happen to know that the process variables follow a Gaussian distribution, which therefore means that the mapping from process variables to read current is nonlinear. The bend in the middle of the plot implies a quadratic mapping. There is a vertical stripe in the lower left; this means that there are region(s) of process variable space with a read current of 0.0, i.e. the bitcell has turned off.

Let us now look at Fig. 3.19 right, which is of sense amp delay. Since this curve is far from linear, it means the delay distribution is far from Gaussian. The three near-vertical stripes mean that there are three modes to the distribution (tri-modal distribution). The jumps between the stripes indicate discontinuities or "cliffs": at these locations, a small step in process-variable space leads to a giant change in performance. If we had taken just 1,000 samples, it would produce just dots in the y-range of about -3 sigma to $+3$ sigma, and we would see just the inner region of the center vertical stripe. That stripe has a slight slope, and would appear linear. Therefore, a sample size of 1,000 might lead us to conclude that the delay distribution is Gaussian. Of course, with more samples we see that it's not Gaussian; the other stripes start at about ± 4 sigma, which means one needs about 1 million MC samples and simulations to start to see those. (In the high-sigma chapter, we will show how to get high-sigma information without needing 1 million simulations).

3.11 Confidence Intervals

In Sect. 3.9, we discussed how mean, standard deviation, and yield are estimated from MC samples. Consider if we took 10 MC samples, and all 10 were feasible. This gives a yield estimate of 100 %. Does that mean that we should trust the design to *really* have 100 % yield? If not, then how do we properly interpret this situation? The answer is in the use of *confidence intervals*: any statistical estimate has a certainty attached to it. That certainty is in the form of an upper bound and a lower bound for a given estimate. For example, if 10/10 samples are feasible, then the upper bound is 100 %, but the lower bound will be 72.2 % (using the Wilson estimate, which we will discuss in more detail later).

As more samples are taken, the width of the confidence interval (CI) will tighten up. For example, verifying that a circuit hits 2-sigma yield with 95 % confidence takes about 80-100 samples. The exact number depends on the actual yield of the circuit. Verifying that a circuit hits 3-sigma yield with 95 % confidence takes about 1400 samples. If the circuit fails to hit the target yield, it will take fewer samples, and substantially fewer samples if the circuit's actual yield is very poor. This is the key to statistical verification, which Chap. 4 will discuss in more detail.

We now explain how to compute CIs for mean, standard deviation, and yield. Say we are given a list of output values $\{x_1, x_2, \ldots, x_N\}$, where N is number of samples. For each statistical measure P (e.g., mean), we estimate the confidence interval by (a) estimating a distribution for the measure P, then (b) computing the upper and lower bound from the distribution. Note that each measure has various ways to estimate the confidence interval. We focus on the most common.

3.11.1 Confidence Interval for Mean (P1)

First, assume that the estimate $P1$ is normally-distributed. This is an acceptable assumption for estimates of mean, from statistics theory. Since a normal distribution is characterized by two parameters (mean and standard deviation), if we compute those two parameters then we will have characterized the distribution of the estimate of mean (P1).

- Estimate of $P1$'s mean = "average" = $\bar{x} = \dfrac{1}{N} \sum_{i=1}^{N} x_i$.

- Estimate   of $P1$'s   standard   deviation = "standard error" = $\dfrac{standard\ deviation\ s}{\sqrt{N}} = \dfrac{\sqrt{\dfrac{1}{N-1} \sum_{i=1}^{N}(x_i - \bar{x})^2}}{\sqrt{N}}$. Note that standard error is *not* standard deviation—a common source of confusion.

Now that we have estimated the two parameters for the distribution of estimate P1, the lower and upper bound for the distribution are simply calculated via statistics on the normal distribution. In particular, to get a CI with 95 % confidence (correct 19 times out of 20), the calculations are:

- Estimate of $P1$'s lower bound = $\bar{x} - 1.96^*$standard_error
- Estimate of $P1$'s upper bound = $\bar{x} + 1.96^*$standard_error

3.11.2 Confidence Interval for Standard Deviation (P2)

Statistics theory dictates that $P2$ follows a χ^2 (Chi-squared) distribution, not a normal distribution. A χ^2 distribution is characterized by the parameters mean and standard deviation. We will show how to calculate $P2$'s mean, then go straight to the confidence intervals.

- Estimate of $P2$'s mean = "standard deviation" = $s = \sqrt{\dfrac{1}{N-1} \sum_{i=1}^{N}(x_i - \bar{x})^2}$.

Here are the estimates for the CI of $P2$. Note how they are best suited to computer-based calculation.

- Estimate of $P2$'s lower bound $= \sqrt{\dfrac{(N-1)s^2}{\chi^2(1-\dfrac{\alpha}{2}, N-1)}}$ where $\chi^2(1-\dfrac{\alpha}{2}, N-1)$ is

the lower critical value of the χ^2 distribution, typically found by numerical computation or by using a lookup table. α is 0.05 for a 95 % confidence level.

- Estimate of $P2$'s upper bound $= \sqrt{\dfrac{(N-1)s^2}{\chi^2(\dfrac{\alpha}{2}, N-1)}}$ where $\chi^2(\dfrac{\alpha}{2}, N-1)$ is the

upper critical value.

3.11.3 Confidence Interval for Yield (P3)

There are several ways to estimate the confidence interval for yield. We describe three approaches. The first, "normal approximation", is simple but makes strong (dangerous) assumptions. The second, "bootstrapped density estimates" uses output margin information for tighter confidence intervals, but carries some extrapolation risk. The third, "Wilson score", is simple and makes no assumptions at all.

Normal approximation interval: This approach is popular because it is simple. It assumes that the binomial pass/fail distribution can be approximated by a Gaussian distribution. This is reasonably accurate when the yield is moderate, but breaks down when yield >99 %. Recall that typical yield targets are 3-sigma (99.86 %) or higher, i.e. yield values where the normal approximation will be inaccurate. This approach is also inaccurate when there is a low number of samples N. Here is how the normal approximation calculates yield CI:

- Yield estimate $(P3) = \dfrac{number\ of\ successes}{number\ of\ samples} = \hat{p} =$ proportion of successes in a

  "Bernoulli trial"

- Estimate of $P3$'s upper bound $= \hat{p} + 1.96\sqrt{\dfrac{\hat{p}\left(1-\hat{p}\right)}{N}}$ (to 95 % confidence)

- Estimate of $P3$'s lower bound $= \hat{p} - 1.96\sqrt{\dfrac{\hat{p}\left(1-\hat{p}\right)}{N}}$ (to 95 % confidence)

Bootstrapped density estimates: In this approach, one makes N_B (e.g. 10,000) estimates of the distribution of the output performance value. Each estimate is made from a slightly different, "bootstrapped", version of the performance output sample data. For each "density estimate" distribution, the yield is computed as described in Sect. 3.8. This results in a list of N_B estimates for yield. The upper and lower bound are simply computed as percentile values from the list.

"Bootstrapping" (Efron and Tibshirani 1994) is a widely used technique in statistics. It can be viewed as a way to take a list of objects, and draw new lists of objects, where each new list can be viewed as a Monte Carlo (MC) sample from a distribution characterized by the original list. To draw a new list, one simply performs sampling with replacement from the original list. From each MC sample of a list, one can perform higher-level calculations (such as our application, which performs density estimation then yield estimation), to get a list of MC sample estimates of yield. While it appears simple, bootstrapping has great value because it enables computation of confidence intervals on a wide range of computational problems where no other techniques are available, or where other techniques make limiting assumptions.

This approach is better than the normal-approximation approach, because it does not assume a normal distribution for pass/fail or for output values. It can give fairly tight confidence intervals on yield, especially when there is high margin between sample values and the spec value. There is actually a pragmatic way to make it work for > 1 performance output value (for details, see Chap. 4 Appendix A).

The biggest drawback of this approach is that it must extrapolate when there is not enough data, and therefore carries the risks associated with extrapolation. We illustrate with an example. Suppose that one has 100 MC samples of an output value. This will have representative sample points out to about 2 sigma. Given that data, this approach will report a confidence interval, and if there is high margin, then the CI could be in the 3 sigma range. This is fine if one is willing to assume that the samples so far adequately represent the distribution, i.e. they can be safely extrapolated. However, if there is a rare failure mode in the distribution, e.g., in 5/1000 samples the performance drops sharply, and none of those samples have been taken yet, then the density estimation will not capture it, and instead returns an overly optimistic CI.

Another drawback is that it takes non-negligible computational effort: for each MC sample of a list, one must make a density estimate, then compute yield from the density estimate. For example, in our experience with $N_B \approx 5000$, kernel density estimation on $\approx 1{,}000$ output values, and yield calculation via bisection search, this requires ≈ 20 s on a single 1-GHz core.

Wilson score interval: This approach (Wilson 1927) was developed as an improvement over the normal approach in computing confidence intervals on a binomial pass/fail distribution. Unlike the normal approach, it has good properties even when the yield is high (>99 %) or when there is a low number of samples N. Unlike the density-estimate approach, it does not assume that extrapolation is acceptable. Therefore, it is the safest of the three approaches described.

- Yield estimate $(P3) = \dfrac{number\ of\ successes}{number\ of\ samples} = \hat{p} =$ proportion of successes in a "Bernoulli trial"

- Estimate of *P3*'s upper bound (+) and lower bound (−), to 95 % confidence, is:

$$\frac{\hat{p} + \dfrac{1}{2N}1.96^2 \pm 1.96\sqrt{\dfrac{\hat{p}\left(1 - \hat{p}\right)}{N} + \dfrac{1.96^2}{4N^2}}}{1 + \dfrac{1}{N}1.96^2}$$

There are other approaches to computing CIs on binomial distributions, such as the Jeffreys CI, the Clopper-Pearson CI, and the Agresti-Coull CI (Agresti and Coull 1998; Ross 2003). Each of those, however, has stability issues in certain cases; whereas the Wilson CI does not.

References

Agresti A, Coull BA (1998) Approximate is better than exact for interval estimation of binomial proportions. Am Stat 52:119–126

Drennan PG, McAndrew CC (2003) Understanding MOSFET mismatch for analog design. IEEE J Solid State Circuits (JSSC) 38(3):450–456

Efron B, Tibshirani R (1994) An introduction to the bootstrap. Chapman & Hall/CRC, London

Hastie T, Tibshirani R, Friedman J (2009) The elements of statistical learning: data mining, inference, and prediction, 2nd edn. Springer, NY

Li X, McAndrew CC, Wu W, Chaudry S, Victory J, Gildenblat G (2010) Statistical modeling with the PSP MOSFET model. IEEE Trans Comput Aided Des Integr Circuits Syst 29(4):599–606

McAndrew CC, Stevanovic I, Li X, Gildenblat G (2010) Extensions to backward propagation of variance for statistical modeling. IEEE Des Test Comput 27(2):36–43

McConaghy T (2010) Symbolic density models of one-in-a-billion statistical tails via importance sampling and genetic programming. In: Riolo R, McConaghy T, Vladislavleva E (eds) Genetic programming theory and practice VIII. Springer, NY (invited paper)

Parzen E (1962) On estimation of a probability density function and mode. Ann Math Stat 33:1065–1076

Rosenblatt M (1956) Remarks on some nonparametric estimates of a density function. Ann Math Stat 27:832–837

Ross TD (2003) Accurate confidence intervals for binomial proportion and Poisson rate estimation. Comput Biol Med 33:509–531

Wilson EB (1927) Probable inference, the law of succession, and statistical inference. J Am Statist Assoc 22:209–212

Chapter 4
3-Sigma Verification and Design

Rapid Design Iterations with Monte Carlo Accuracy

Abstract This chapter explores how to efficiently design circuits accounting for statistical process variation, with target yields of two to three sigma (95–99.86 %). This yield range is appropriate for typical analog, RF, and I/O circuits. This chapter reviews various design flows to handle statistical process variations, and compares these flows in terms of speed and accuracy. It shows how a sigma-driven corner flow has excellent speed and accuracy characteristics. It then describes the key algorithms needed to enable the sigma-driven corner flow, namely sigma-driven corner extraction and confidence-based statistical verification. Some enabling technologies include Monte Carlo, Optimal Spread Sampling, confidence intervals, and 3σ corner extraction.

4.1 Introduction

Chapter 2 addressed PVT variation, and its specific use where Fast, Typical, and Slow (F/T/S) process corners could adequately model the effects of process variation. These corners represent statistical bounds of global process variation for the device-level performances of speed and power. The PVT approach to modeling process variation is appropriate when the effects of local process variation ("mismatch") is minimal, and when the device-level performances of speed and power directly relate to all circuit-level performances, such as digital standard cell performances of speed and power. The PVT approach is also used when accurate statistical models of process variation are not available. When these conditions are not met (i.e. when device speed/power don't directly relate to circuit output measures of interest, or when local variation is significant), then a statistical approach is more appropriate.

T. McConaghy et al., *Variation-Aware Design of Custom Integrated Circuits: A Hands-on Field Guide*, DOI: 10.1007/978-1-4614-2269-3_4,
© Springer Science+Business Media New York 2013

This chapter is concerned with designing circuits when global or local process variations need to be handled statistically and the target yield is 2–3 sigma, i.e. 95–99.86 % yield. This includes analog, RF, and I/O circuits. Such circuits can have up to thousands of devices. Yield numbers of 2–3 sigma imply about 80–1,400 Monte Carlo (MC) samples are required for analysis. Since the target is typically 3-sigma, we will often just specify "3-sigma" when applying to other sigma values.

Whereas this chapter is concerned with 2–3 sigma statistical design, Chap. 5 discusses *high-sigma* (3.5–6 sigma) design. High-sigma design is applicable to circuits that are replicated many times on a chip. This includes memory building blocks such as bitcells and sense amps, and digital standard cells. Because they are replicated and their functionality is simpler, high-sigma circuits tend to be smaller, typically with 4–50 devices. Yield numbers of 3.5–6 sigma imply about 50,000–5 billion Monte Carlo samples for analysis. High-sigma design is also applicable where failure (even rare failure) is catastrophic, such as in some automotive IC applications and medical applications.

Along the way, this chapter will resolve such designer questions as: How many MC samples do I need? If I run 100 MC samples and they all pass, have I verified to 3σ? And most importantly: is there a way to design with rapid design iterations, yet as accurately as if I were doing MC sampling at each iteration?

This chapter is organized as follows. First, it reviews various design flows handling 3-sigma statistical variation and describes how the sigma-driven corners flow gives the best combination of speed and accuracy. Then, it describes the components that enable the sigma-driven corners flow: sigma-driven corner extraction, and confidence-driven 3σ verification. It then describes Optimal Spread Sampling, which has lower variance than pseudo-random or Latin Hypercube sampling approaches. Finally, it provides an example case study comparing design of a D-flip-flop using four different variation-handling approaches, applying the learnings from the rest of the chapter.

This chapter also includes an appendix on density-based yield estimation on more than one output, and another appendix containing details of low discrepancy sampling.

4.2 Review of Flows that Handle 3-Sigma Statistical Variation

4.2.1 Introduction

This section reviews the flows handling 3-sigma statistical variation, comparing them in terms of speed and accuracy. The flows range from simple PVT flows and direct MC flows to more complex flows like worst-case distances and response surface modeling. In the flows, the actual design iterations may be manual or

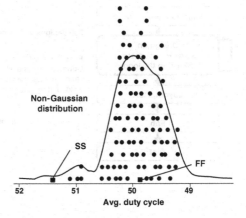

Fig. 4.1 FF/SS corners versus distribution, for the average duty cycle output of a PLL VCO, on GF 28 nm. Adapted from (Yao 2012)

otherwise (Chap. 6 discusses the options in detail). The final flow is a sigma-driven corner flow, which as we will show, has the best tradeoff between speed and accuracy.

Because the flows include changing the design variables, we need a way to compare different approaches fairly, independent of designer skill and level of designer-knowledge about the circuit.[1] We do this with a simple assumption: in the design loop, the designer will consider 200 designs (for all flows). Where applicable, we assume that there are two performance specifications. Finally, where applicable, we assume that there are 200 devices, and 10 local process variables per device, as is common with the backward propagation of variance formulation of process variation.

To better analyze certain flows, we will use the representative performance distribution shown in Fig. 4.1. It is the distribution of average duty cycle, of a PLL VCO, on Global Foundries' 28 nm process. We note immediately that the distribution is not bell-shaped, and therefore it is not Gaussian.

4.2.2 Flow: PVT (with SPICE)

This flow, shown in Fig. 4.2 left, is representative of the variety of PVT flows discussed in the PVT chapter. In the first step, the designer selects the topology and performs initial sizing. Then, the designer improves the circuit against a set of PVT corners, using SPICE simulation for feedback about circuit performance. The PVT corners may have been set by the user, or extracted using a Fast PVT approach. Once the designer is satisfied with the design's performance, he proceeds to layout, RC extraction, final verification, tape-out, fabrication, and test.

---

[1] Or, if the designer is applying an automated sizer, we want to be independent of the competence of the sizer.

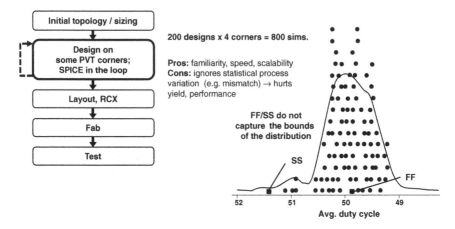

Fig. 4.2 Flow: PVT (with SPICE)

This flow's advantages include familiarity, speed, and scalability. It is familiar because most designers have worked with some PVT-style flow in their careers. It is fast because it only needs one simulation for each PVT corner, at each candidate design. Our example has 4 corners, which gives us a total of 200*4 = 800 simulations. The flow is scalable because the number of simulations is independent of the size of the circuit: regardless of whether the circuit has 10 or 10,000 devices, it requires the same number of simulations to analyze the design.

However, the flow does not accurately model variation. It completely ignores local process variation, which is a dominant factor in many circuits, analog and otherwise. It also does a poor job of approximating global process variation: the F/T/S corners are only representative of performance bounds if the circuit's target performances correlate with device-level speed and power performance. This is usually not true for analog circuits, and not true for many other circuits as well. Figure 4.2 right illustrates how the FF/SS corners do not capture the bounds of the distribution in the example PLL VCO circuit. Because the flow does not accurately model variation, the resulting design could suffer significant yield loss.

4.2.3 Flow: PVT ± 3-Stddev Monte Carlo Verify

Figure 4.3 left shows this flow. The key steps are in bold. First, the user designs against the PVT corners, like in the last flow. Then, in order to measure the effect of statistical process variation, the user runs a Monte Carlo (MC) sampling. He runs just 100 MC samples, then estimates mean (*mean*) and standard deviation (*stddev*) of the performance. Then, he measures the performance value at *mean* − 3*stddev and/or *mean* + 3 * *stddev*. If those performance values are within the specifications, then the designer is satisfied.

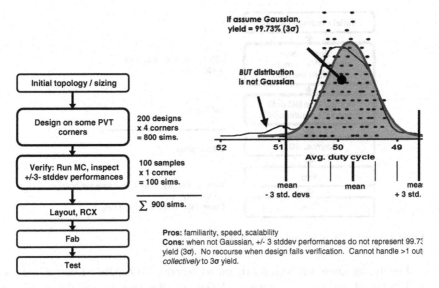

Fig. 4.3 Flow: PVT ± 3-stddev MC verification

Figure 4.3 top right illustrates what is implicitly going on. When *mean −
3*stddev* and *mean + 3 * stddev* are within specification bounds, it means yield
has met the target of 3-sigma (99.73 %), *if* the distribution is Gaussian. As we
discuss shortly, this is a very big "if".

This flow has the advantages of the previous flow: familiarity, speed, and
scalability. In addition, its verification step is more accurate, because it accounts
for local process variation, and for global process variation better. It is familiar,
because many designers design against PVT corners, then verify by running MC.
It is fast because both steps are fast: the PVT step is on a small set of corners, and
MC verification only needs 100 samples. It is scalable with the number of devices
and variables because both the PVT and MC steps are scalable.

However, this flow has several issues. First, it assumes that the distribution is
Gaussian. In Fig. 4.3 top right, we show samples from the example PLL VCO
circuit. On those samples, we overlay a Gaussian distribution (bold line), and a
more accurate non-Gaussian distribution estimated from the data.[2] We see that the
distributions are quite different. In the context of the flow, this can lead to over-
optimism—the user may conclude that the design is fine when it is not.

Another issue is in fixing the design. When the design fails verification, there is
no obvious next action to perform, to improve the design.

---

[2] To be precise, the non-Gaussian distribution is estimated with Kernel Density Estimation
(KDE).

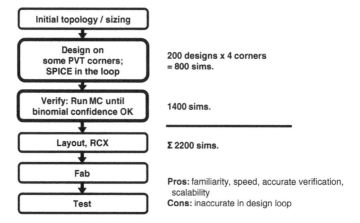

Fig. 4.4 Flow: PVT + binomial MC verification

Finally, the approach does poorly on >1 outputs. To handle >1 output, the user will test *each* output on 3 standard deviations. But that ignores the correlation among outputs. When two outputs have weak correlation, different sample points fail for different outputs, and the failure rate is approximately double what this method might predict.

4.2.4 Flow: PVT+ Binomial Monte Carlo Verify

Figure 4.4 illustrates this flow. The key steps are in bold. First, the user designs against the PVT corners, like in the last flow. Then, in order to measure the effect of statistical process variation, the user runs a Monte Carlo (MC) sampling. Whereas the last flow stopped at 100 samples and assumed Gaussian, this flow uses an assumption-free binomial pass/fail distribution, and MC sampling only stops sampling once statistically confident (according to the distribution) that the design has passed or failed the target yield. Section 4.4 elaborates on this approach to statistical verification.

Like the previous flow, this flow is fast, familiar, scalable, and accurate in the sense it acknowledges local and global statistical process variation. Its verification is even more accurate,[3] because it does not assume that the performance distribution is Gaussian. It is fast because both steps are fast: the PVT step is on a small set of corners, and MC verification only needs $\approx 1,400$ samples to verify to 3-sigma with 95 % statistical confidence. It is scalable with the number of devices and variables because both the PVT and MC steps are scalable. MC accuracy only depends on the number of samples, but is fully independent of the number of

---

[3] It is accurate to the extent that the statistical MOS models are accurate.

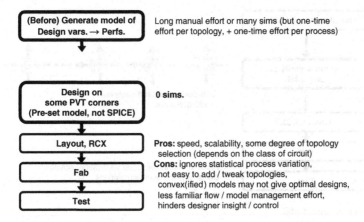

Fig. 4.5 Flow: PVT with convex models

variables. MC always needs 1,400–5,000 samples for accurate 3-σ verification, regardless of whether a circuit has 10 or 10,000 devices. Finally, unlike PVT-only flow, this flow includes accurate statistical verification due to its MC-sampling step.

The main disadvantage of this flow occurs when the MC-based verification fails to meet the target yield. As shown, this flow only has a PVT-based design loop, which does not have Monte Carlo accuracy. It is possible to treat the worst-case MC samples as corners, then design against them. However, these MC corners do not imply any particular yield value, which means the user has no visibility into the effect on yield as he improves the design against the MC corners. For example, a worst-case MC sample may implicitly[4] correspond to a target of 80 % yield. If the designer solves for that corner, the yield will be 80 %. Section 4.3.2 demonstrates this issue on several benchmark circuits.

4.2.5 Flow: PVT with Convex Models

Figure 4.5 illustrates this flow. The idea is for an expert in modeling to pre-generate models that map design variables to performance. Then, at design time, the designer runs an automated sizing "optimization" on the models to quickly find the design that meets the target performances. The models are convex or convexified and look like one big hill, so that even a fast hill climbing-style optimizer can find the global optimum. For analog circuits, the convex-optimization approach was first published in (Hershenson et al. 1998), and has evolved

---

[4] We say "implicitly" because the designer has no clear way to see the sample's relation to yield.

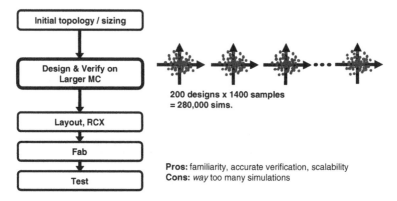

Fig. 4.6 Flow: direct MC

over the years to its current commercial incarnation (Magma Design Automation 2012).

The detailed flow is as follows. For each new process node or topology, a modeling expert creates a convex model (e.g. a polynomial) mapping design variables to worst-case performance across PVT corners. This is a semi-automated flow; it typically requires the expert to have knowledge about the circuit behavior and how to write analytical performance equations for it. The model is made available for use by designers, within the context of the convex optimization tool. Savvy designers can also input their own models.

The next steps are the designer's work flow. There is no need for topology selection and initial sizing. Instead, the user inputs the circuit type, process node, and performance targets. Convex optimization is run on the convex model of each model in that circuit type and process node. The tool outputs a sized net list for the circuit topology that best meets the performance targets. Optimization takes seconds for small circuits, or minutes to hours for very large circuits. Finally, the usual steps of layout, fab, and test are performed.

The great advantage of this flow is its speed at design time: it takes no simulations. It also scales well, as demonstrated on large ADCs, PLLs, and more. Finally, the flow also bypasses the need for the designer to do topology selection and initial sizing, which can be time-consuming.

This flow has several disadvantages. First, by only working on PVT corners, the model of process variation is inaccurate. Second, it is not easy to add or even adjust topologies because it requires manual creation of new models for each new topology or topology change. Third, the designs returned may not be optimal even for just PVT variation because the convex models themselves can be inaccurate; though in practice the numbers come out surprisingly close if one does a careful job of modeling. This flow is a large departure from the familiar design flow, and gives up much designer insight and control; both characteristics make this flow less palatable to many designers.

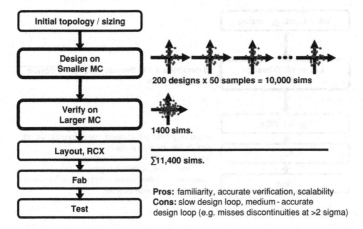

Fig. 4.7 Flow: light + heavy direct MC

4.2.6 Flow: Direct Monte Carlo

This flow, shown in Fig. 4.6, is a baseline flow that illustrates an extreme of good accuracy but poor speed. In the design and verify step, shown in bold, the designer runs full MC verification on every candidate design. For example, it typically takes $\approx 1{,}400$ samples and simulations to achieve 3σ accuracy. Given 200 design iterations, this amounts to 280,000 simulations.

The direct MC flow actually has several advantages: it is very familiar and easy to implement, it is accurate to 3σ in both the design loop and in verification, and it is scalable on circuit size due to the scalability of MC with dimensionality. However, its fatal disadvantage is that it is simply too slow due to the large number of simulations required.

4.2.7 Flow: Light + Heavy Direct Monte Carlo

Figure 4.7 illustrates this flow. It is a more pragmatic variant of the previous MC-based flow. In the design loop, it runs just a small number of MC samples (e.g. 50 samples, good for nearly 2 sigma) on each design candidate. Once the designer is satisfied with the performance and yield on the 50 MC samples, he verifies on a larger MC run of 1,400 samples for 3σ accuracy.

The advantages of this approach are: familiarity, statistically-aware design iterations (but to <2 sigma), 3σ-accurate verification, and scalability. But it has significant disadvantages. First, inside the design loop, 50 simulations per design

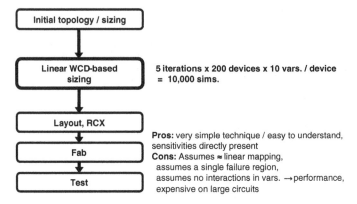

Fig. 4.8 Flow: linear worst-case distances

leads to fairly slow iterations, and high overall simulation cost. Second, the design loop is only accurate to 2 sigma. For example if performance dramatically drops off at 2.5 sigma, the flow will miss it during design, leading to an expensive iteration with design and verification.

4.2.8 Flow: Linear Worst-Case Distances

This flow, shown in Fig. 4.8, is based on Schenkel et al. (2001). One iteration of linear Worst-Case Distances (WCD)-based sizing works as follows. The circuit is simulated at the design point and nominal process values. Then, each process variable and design variable is perturbed one at a time; at each perturbation, the circuit is simulated. From the data at nominal and at perturbations, a linear model is constructed. The linear model maps design variables and process variables to output performances. Using the model, a new design point is chosen: it is the point that maximizes the yield according to the linear model.

For an typical example circuit having 200 devices and 10 process variables per device, the number of local process variables is much larger than the number of global process variables or design variables. So, roughly speaking, each iteration takes 200 devices × 10 variables per device, or 2,000 simulations. According to the literature, linear WCD typically needs about 5 iterations, for a total of 10,000 simulations in this example.

The advantages of this approach are its simplicity and ease of understanding. As a bonus, it makes sensitivities of output performances to design variables and process variables directly available. The major disadvantage is inaccuracy due to the following three assumptions: an approximately linear mapping, a single region where specifications fail, and weak or no interactions among variables in the

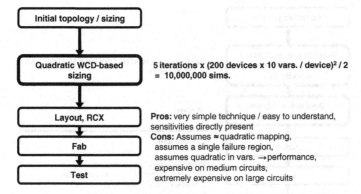

Fig. 4.9   Flow: quadratic worst-case distances

mapping to performance. While the approach is very fast on small circuits with few process variables per device, it gets expensive on medium-sized and larger circuits because it requires one simulation for each process variable of each device.

4.2.9 Flow: Quadratic Worst-Case Distances

This flow, shown in Fig. 4.9, is quite similar to the linear Worst-Case Distances (WCD) approach of the previous section. The difference is that each iteration constructs a quadratic model rather than a linear model (Graeb 2007). Because it does quadratic modeling in a sequential fashion, it can be viewed as a variant of sequential quadratic programming (SQP) (Boggs and Tolle 1995).

Quadratic WCD is more accurate than linear WCD because a quadratic mapping covers a broader range of circuit outputs. However, it takes $1 + n(n-1)/2$ simulations per iteration rather than $1 + n$ simulations, where n is number of variables. This dramatically affects the scalability of the approach: for a design with 2,000 process variables and 5 iterations, it needs 10 million simulations. Furthermore, the quadratic mapping is still not accurate enough for many circuits. For example, quadratic mappings do a poor job of handling discontinuities. A discontinuity might occur, for example, when a particular combination of process variation values causes a transistor to shut off, which in turn makes circuit performance drop sharply.

4.2.10 Flow: Response Surface Modeling

Figure 4.10 illustrates the response surface modeling (RSM) flow. In the first step, the user selects the initial topology and sizes as usual. Then, a tool automatically constructs some form of response surface model, either quadratic or more general,

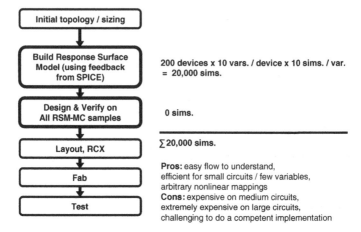

Fig. 4.10 Flow: response surface modeling (RSM)

using feedback from SPICE. After that, all analysis is performed on the model, which evaluates orders of magnitude faster than SPICE.

The automatic model construction usually has an initial phase and an adaptive phase. The initial phase starts with some form of well-spread sampling or design of experiments (DOE) in the space of design and process variables, and then simulates each of those sampled points. Each iteration of the adaptive phase builds a model using all samples and simulations so far, chooses one or more new design and process points based on feedback from the model, then simulates the chosen points. Models may be linear, quadratic, splines, or may incorporate more advanced approaches such as feed forward neural networks, support vector machines (SVMs) (Cortes and Vapnik 1995), Gaussian process models (GPMs) (Cressie 1989), or fast function extraction (FFX) (McConaghy 2011). Each modeling approach has its own advantages and disadvantages in terms of build time, model accuracy, scalability to number of dimensions, and scalability to number of samples.

In our experience and from the literature, if one has a sufficiently accurate and sufficiently nonlinear modeling approach, as well as a competent adaptive sampling algorithm, then to get a model with sufficiently accurate predictive abilities takes at least 10 × the number of input variables. Therefore, for a circuit with 200 devices and 10 process variables per device, or about 2,000 input variables total, the approach will need about 2,000 × 10 = 20,000 simulations to build a sufficiently accurate model. A circuit with 20 devices will need 2,000 simulations, and a circuit with 2,000 devices will need 2,00,000 simulations.

The advantages of this approach include ease of understanding, high efficiency on very small circuits, and, assuming a competent model and adaptive sampling algorithm, good accuracy. The main disadvantage is its high to extremely high simulation cost on moderate-sized and large circuits. Furthermore, developing a competent model and adaptive sampling algorithm requires a high degree of expertise.

Fig. 4.11 The idea of sigma-driven corners

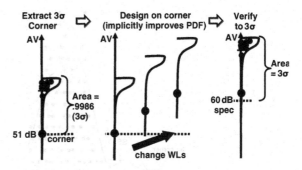

4.2.11 Flow: Sigma-Driven Corners

Figure 4.11 shows the idea behind the sigma-driven flow and its three core steps. In the first step, the designer extracts a corner for each output (e.g. gain AV), where the corner is the 99.86th lower bound percentile of the AV distribution. This is a *sigma-driven* corner. In the second step, the designer changes the circuit's sizings with feedback from SPICE, simulating at the sigma-driven corner until meeting the specification(s). Note in the figure that as the corner's AV increases, the whole delay distribution is implicitly pushed upward without having to simulate the whole distribution. In the final step, the designer verifies the yield to see whether the yield is $\geq$99.86 % within a statistical confidence limit (e.g. 95 % confidence). If the yield passes, the designer is done; otherwise, he extracts a new AV corner and re-loops.

Naturally, one may handle more than one specification by simply having one corner for each specification. When corners are extracted, the overall yield must meet the target sigma, and each of the partial yields needs to meet or exceed the target sigma.

The sigma-driven design flow overcomes the issue where yield estimation within the sizing loop is prohibitively expensive. Each sizing candidate within the loop has just a handful of simulations on corners, rather than a full Monte Carlo sampling.

Figure 4.12 shows an example of 3σ corners extracted for the average duty cycle output of a PLL VCO on a GF 28 nm process. Note how the 3σ corners accurately capture the upper and lower statistical bounds of the output's distribution, unlike the traditional FF/SS corners.

Figure 4.13 shows the sigma-driven corners concept in the context of an overall design flow. Following the step of topology selection and initial sizing, there are three core steps: extracting the 3σ corners, designing on the corners, and verifying to 3σ yield. It turns out that enhanced versions of Monte Carlo are useful algorithms for extracting 3σ corners, and verifying to 3σ yield in a fast, accurate, and scalable fashion.

Usually, circuits will pass the 3σ verification step because the sigma-driven corners are extracted with good accuracy. This is the "1 design pass" flow in the

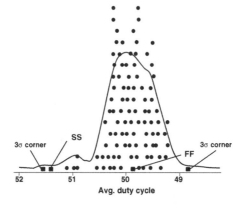

Fig. 4.12 3σ corners versus SS/FF corners for the average duty cycle distribution for a PLL VCO on a GF 28 nm process. Adapted from (Yao 2012)

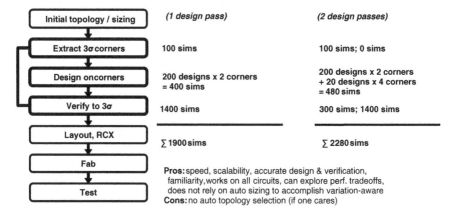

Fig. 4.13 Flow: sigma-driven corners, with 3-sigma yield target

center column of Fig. 4.13. Extracting the corners takes about 100 simulations, design takes about 400 simulations, and verification takes about 1,400 simulations, for a total of 1,900 simulations.

However, if there is significant interaction between process variables and design variables in the mapping to output performances, then the more the design changes, the more inaccurate the corners become, and the greater the risk the circuit does not pass the 3σ verification step. If this happens, then a second iteration is needed where new 3σ corners are extracted, the design is tuned, though more locally this time, against corners from both iterations, and finally verified to 3σ yield. The total number of simulations remains modest at 2,280 simulations. Two rounds have only marginally more simulations than one round because (1) the first round of verification stopped as soon as 3σ verification failed (e.g. after 300 simulations); (2) corner extraction for the second round is basically free because it reuses the first round's verification simulations; and (3) the second round has fewer design changes.

We can perform a similar flow for a 2-sigma (95 % yield) target, with even fewer simulations. Extracting 2σ corners takes 50 simulations, and verifying to 2σ with 95 % statistical confidence takes ≈ 80 simulations.

The sigma-driven corners flow has many advantages:

- **Fast:** It is fast because each of its steps is fast: corner extraction needs just ≈ 100 simulations, design iterations are on just a small set of corners (one per spec), and verification stops as soon as it is statistically demonstrated that the design has met or failed to meet the target yield.
- **Accurate:** The flow is accurate to 3σ in both design and verification because the corners are 3σ accurate and the verification step is accurate to 3σ. It uses SPICE-in-the-loop, and does not make simplifying approximations.
- **Scalable:** The flow is scalable because each step is scalable: corner extraction and verification are based on MC and are independent of circuit size, and number of corners in the sizing step is also independent of circuit size.
- **Easy to adopt and familiar:** It uses a corner-based flow, just like the often-used PVT corners flow. Only now, the corners are *really* good—they are finally accurate to analog circuit performance boundaries rather than reflective of digital device performances.
- **Flexible:** This flow works on all circuits; it does not rely on an approximately linear or quadratic mapping. It is flexible another way: it can be used with either automated or manual sizing flows. Put another way: contrary to some of the literature, one does not need to resort to automated design to do statistically accurate variation-aware sizing.
- **Tradeoff exploration:** This flow does not need any circuit specifications at all because it extracts corners on the bounds of performance distributions. This means that designers can explore tradeoffs among performance outputs, subject to the target yield that the corners were extracted at.

The sigma-driven flow resolves designer questions as follows.

- **Question:** How many MC samples do I need? *Answer:* If you are doing verification, then *adaptively* decide when to stop, based on statistical confidence that the design meets or does not meet the target yield. If you are doing corner extraction, then ≈ 100 MC samples is about right.[5]
- **Question:** If I run 100 MC samples and they all pass, have I verified to 3σ? *Answer:* No. The key to answer this is with *confidence intervals*: given the number of passes and number of MC samples, we can compute a lower and upper bound for what the yield will be with 95 % confidence. Typically, you need $\approx 1,400$ MC samples to verify that a circuit has passed 3σ yield; more if it is very close to precisely 3σ yield, and fewer if it fails 3σ yield.
- **Question:** Is there a way to design with rapid design iterations, yet as accurate as if I was doing MC sampling for each design? *Answer:* Yes, using the sigma-driven corners flow.

---

[5] Benchmarks in Sect. 4.3.2 elaborate on this statement.

Fig. 4.14 Flow: summary of
flows for 3-sigma statistical
design

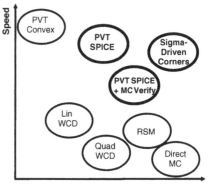

4.2.12 Summary of Flows for Three-Sigma Statistical Design

Figure 4.14 summarizes the flows we just discussed in terms of speed and accuracy. The fastest (top left) is PVT convex since it does not use any simulations during the design phase. However, it is quite inaccurate. The slowest (bottom right) is direct MC, because it requires >1,000 simulations for each design candidate. On medium- and large-sized circuits, the WCD variants and RSM are quite slow. In the far top right, in bold, is the sigma-driven corners flow. It is as accurate as a direct MC flow, but very fast because it has corners in the iterative design loop. We have also bolded the two PVT SPICE variants, because they have uses in some contexts, for example when global variation and power and speed measures matter the most.

The rest of this chapter focuses on the sigma-driven corners flow. This flow has two key analysis steps: 3σ sigma-driven corner extraction, and 3σ verification. We describe them in that order.

4.3 Sigma-Driven Corner Extraction

Sigma-driven corner extraction is a crucial element in the sigma-driven design flow of Fig. 4.11. Sigma-driven corner extraction takes in a sized circuit and a target yield, and outputs one or more corners, where each corner is a point in process variation space. This section describes the general idea of corner extraction, and the specific approach we take.

In corner extraction, we aim to find a corner that represents the bounds of performance. Since the performance is a distribution that typically has nonzero density values extending from $-\infty$ to $+\infty$, a minimum and maximum bound would be $-\infty$ and $+\infty$, which is not meaningful. Instead, we are concerned with a

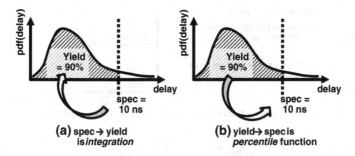

Fig. 4.15 Relation between yield and output specification for a 1-dimensional output case

corner that represents a *statistical* bound: we want a corner at a 3σ target yield, such that if we were to measure the design's yield (e.g. with MC), the yield will come out as 3σ.

Remember, these are for corners at the *circuit* level, not the device level. Because they are specific to the circuit at hand, they cannot be pre-computed and shipped in the PDK the way that device-level corners are.

There has been a small amount of research on circuit-level corner extraction. (Silva et al. 2007) extracted the delay corner on digital cells, but it was restricted to the delay output and specific topology characteristics of digital cells, and assumed a linear mapping from process variables to output. (Zhang et al. 2009) makes a quadratic mapping from process variables to output, which naturally restricts its accuracy to circuits that can be modeled quadratically. It also relies on constructing an accurate model, which strongly depends on the number of process variables. Finally, there is the practice of extracting 3-sigma corners for *devices*, bounding the device speed and power (F and S model-sets).

4.3.1 Sigma-Driven Corner Extraction with 1 Output

4.3.1.1 Review: Yield $\leftrightarrow$ Specification

This section reviews a building block of corner extraction: the functions that relate yield and specification value to the probability density function (PDF) of an output. Figure 4.15a shows that we can compute a yield given a target spec. It computes via *integration*: yield is the area under the PDF in the region where the spec is met. Figure 4.15b goes the opposite direction, computing a spec given a target yield. The opposite of integration is the *percentile function*. For example, if we want the spec given a yield of 90 %, then we compute the output value at the 90th percentile.

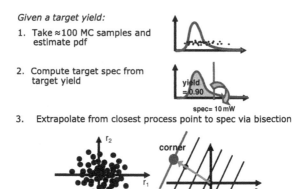

Given a target yield:

1. Take ≈100 MC samples and estimate pdf

2. Compute target spec from target yield

3. Extrapolate from closest process point to spec via bisection

Fig. 4.16 Algorithm for sigma-driven corner extraction with 1 output

In the context of corner extraction, we will be using the latter flow, going from a target yield to a target spec.

4.3.1.2 Sigma-Driven Corner Extraction: Algorithm

Figure 4.11 illustrated the overall flow for sigma-driven design. Its left figure also started to hint at how 3-sigma corner extraction is done. The general idea is to take some MC samples, estimate an output PDF, find a target spec via the percentile function, and finally find a process point that has the target spec value.

Figure 4.16 shows details of the algorithm for sigma-driven corner extraction at a fixed design point. It inputs a target yield, such as 3σ (99.86 % yield).

The first step simulates 100 MC samples to get 100 output values, then estimates a PDF from them using kernel density estimation (KDE) or another density estimation technique.

The second step uses the percentile function to compute a target output specification value from the target yield. Where analytical percentile functions are not available for the PDF, bisection search is a reasonable approach to compute the percentile function to arbitrary accuracy.

The third step finds a process point that gives the target output specification. The algorithm does this by (1) finding the MC sampled process point that gives an output value closest to the target, then (2) performing one-dimensional search between that process point and the nominal process point, using SPICE-in-the-loop to measure output values, until the simulated output value is within a target tolerance of the target spec value. The one-dimensional search starts as a trust region method. If that fails, then a bisection search is employed. If that fails, then more MC samples are taken, and the one-dimensional search is restarted.

The algorithm is deceptively simple, fast, and scalable because it decomposes the problem in an elegant fashion.

To illustrate, consider if Steps 2 and 3 were treated as a single problem. It would search for an n-dimensional process point with maximum probability density, subject to meeting the target yield. Each candidate process point is simulated to give a candidate spec value, from which candidate yield is computed. Put another way, it performs optimization in n-dimensional process variable space with SPICE-in-the-loop.

The single-problem approach is computationally expensive because it must perform SPICE-in-the-loop optimization in n-dimensional process variable space. In contrast, by decomposing it into two separate steps (step 2 and 3), we only need SPICE-in-the-loop optimization on one variable. In our experiments, we found that recasting this decomposition-based approach had negligible effect on the quality of the results, yet had far faster runtime.

The target specification values used in step 2 and step 3 are *not* to be confused with actual circuit specifications. They are simply the value of the output at the 3-sigma mark, as an intermediate step to find a corner process point at the output's 3-sigma mark. The target specification can be ignored once the corner process point is found.

The algorithm is fast, accurate, and scalable. It is fast because each of the three steps is fast: step 1 needs just 100 samples (simulations), step 2 is nearly instantaneous because no simulations are needed, and step 3 typically needs just 5–20 simulations per output because it is doing a simple 1-d search. It is accurate because each step is accurate, always using SPICE for feedback and avoiding making simplifying assumptions. It is scalable to large circuits because each step is scalable: step 1's MC is independent of input dimensionality, and step 1's density estimation is just one dimension; step 2 only has one dimension to compute the percentile on; and step 3 has only one dimension for its SPICE-in-the-loop optimization.

These properties allow corner extraction to work well even for extremely large circuits, having thousands of devices or more.

4.3.2 Benchmark Results on Sigma-Driven Corner Extraction

This section assesses the accuracy of sigma-driven corner extraction. The main research questions are:

- How accurate is sigma-driven corner extraction (CX), compared to (1) simply picking the worst-case minimum or maximum from MC samples (WC) and (2) compared to a target yield?
- How much does the number of MC samples affect accuracy in CX (and WC)?

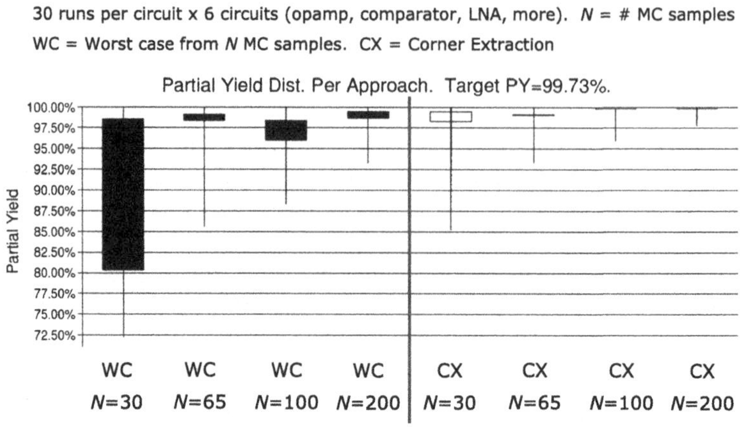

30 runs per circuit x 6 circuits (opamp, comparator, LNA, more). *N* = # MC samples
WC = Worst case from *N* MC samples. CX = Corner Extraction

Fig. 4.17 Benchmark results on sigma-driven corner extraction

The experimental setup used to answer these questions is as follows. We had six analog circuits: an opamp, a comparator, a low noise amplifier, and three more. For each circuit, we compared two approaches: WC and CX. For each approach, we compared four different numbers of MC samples: 30, 65, 100, and 200. For each number of MC samples, we did 30 runs of both MC sampling and corner extraction.

We established accurate reference values for the distribution by pooling together all the MC samples for a given output and computing a density estimate. We set our target yield as 99.73 %, which is 3σ using a two-tailed assumption.

Figure 4.17 shows benchmark results for each combination of approach and number of MC samples. For example, the box plot[6] on the far left is for the WC approach on 30 MC samples. For each circuit, it did 30 runs. On each run, it chose the statistical process point that caused worst-case performance to be a corner, and the corresponding corner's output performance value. It then computes the yield of this output performance value on the reference PDF. Recall that since the aim of the corner is to be at the 99.73[th] yield percentile, the ideal corner output performance value would give an output yield of 99.73 %. However, we see in the box plot on the far left that the values range from 72.5 % up to 100.0 %, with most values between 80.0 and 98.0 %. This means that individual runs gave values like 77.5, 82.5 %, etc. This is a big problem for users of the WC approach for corner extraction: the designer is hoping for a yield ≈ 99.73 % when the corner is solved, but when the corner actually gets solved, the yield is only 77.5, 82.5 %, etc.

If we increase the number of WC samples from 30 to 65 (Fig. 4.17, going from the box plot on the far left, one to the right), we see that the box plot tightens up

---

[6] A box plot is another way of visually representing a distributed set of sample data. The box contains 50 % of the samples, from the 25th to the 75th percentile. The lines extending from the box go to the maximum and minimum sample values seen.

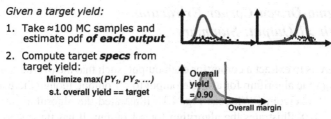

Given a target yield:

1. Take ≈100 MC samples and estimate pdf *of each output*

2. Compute target *specs* from target yield:

   Minimize max(*PY₁, PY₂, ...*)
   s.t. overall yield == target

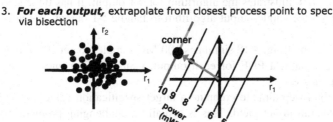

3. *For each output,* extrapolate from closest process point to spec via bisection

Fig. 4.18 Algorithm for sigma-driven corner extraction, with >1 output

substantially. Then, going from 65 to 100 samples, the box plot loosens, then going from 100 to 200 samples it tightens up more In general, the WC box plots are quite scattered. This should not be surprising: the so-called corners are not extracted in a way to be aware of yield at all. But the resulting problem is that when users design against these statistical corners, they have no idea of the effect on yield.

Let us now examine the results for sigma-driven corner extraction (CX) in the four box plots on Fig. 4.17 right. At N = 30 samples, most corners will result in yields of >98 %. We see that as the number of MC samples increases from N = 30–65, then 100, then 200 samples, the distribution tightens up significantly. By the time it hits 100 MC samples, it is already near-perfect and aligning very well with the target of 99.73 %, which is why we recommend that 3σ corner extraction should typically use ≈100 MC samples.

We can also compare the accuracy of WC to CX. We can compare them at the same number of samples, such as WC N = 30 versus CX N = 30. Whereas yields of corners for WC N = 30 mostly vary from 80.0 to 98.0 %, the yields for CX N = 30 mostly vary between 98.0 and 99.5 %, a much tighter distribution. In fact, WC N = 200 is about comparable to CX N = 30, a 200/30 = 6.6× speedup. Furthermore, as WC gets more samples, it will not converge because it never has a target yield, whereas CX will be able to take further advantage of the increased amount of data.

This concludes the benchmarking of sigma-driven corner extraction. We have demonstrated how sigma-driven corner extraction returns accurate corners, and how using worst-case MC samples does not.

4.3.3 Sigma-Driven Corner Extraction with >1 Output: Summary

The aim here is to extract a corner for each output, such that the *overall* yield is 3σ. We can adapt the algorithm for a single output to >1 output, though there are a few challenges to overcome. Whereas Fig. 4.16 illustrated the algorithm for a single output, Fig. 4.18 illustrates the algorithm for >1 output. It has three steps, each a revision from the single-output algorithm to handle >1 output. We now describe the steps.

The first step of Fig. 4.18 simulates 100 MC samples to get 100 values *for each output*. It estimates a PDF for each output, using KDE or another density estimation technique.

The second step aims to compute a target specification value *for each output*, given an overall target yield. This is actually a challenging problem, so we will elaborate below.

The third step finds a process point that gives the target specification value *for each output*. It uses the same 1-d search techniques as the single-output case described earlier.

We now elaborate on the second step, which inputs a target yield and returns a target spec for each performance output. This is actually an underdetermined problem because it has more degrees of freedom (target spec values) than fixed values (single target yield). To address that, we can cast this step as an *optimization* problem, where we minimize *some* objective, while meeting the constraint that the final chosen specs give a yield equal to the target yield. The optimization space is possible spec values.

In deciding what objective to use, let us consider the following. First, for all but one output, pick very loose output spec values that lead to extremely high partial yields (e.g. 6σ). Then, compute the remaining output's spec such that its partial yield is equal to the target yield. The only output that will affect overall yield is the remaining output, which means the target yield constraint is met. However, this is a very poor solution, because most outputs will have extreme-valued output target values, which will lead to very extreme-valued corner process points; those extreme corners will have very poor performance on SPICE.

This thought exercise gives us intuition about what objective might be suitable: we want to ensure that it is not too hard to design against any of the outputs' corners. Put another way, all outputs' corners should be balanced in terms of difficulty. Since higher partial yields (PYs) lead to higher difficulty, we can translate this insight into an objective: minimize the maximum of outputs' PYs. With this objective in hand, we have determined the overall problem formulation of step 2, as shown in Fig. 4.18. We can solve the constrained optimization problem with a competent off-the-shelf nonlinear optimizer such as an evolutionary algorithm (EA). Optimization is relatively cheap because each evaluation is cheap, as SPICE is not in-the-loop.

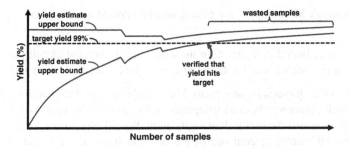

Fig. 4.19 The confidence interval (CI) width tightens as more Monte Carlo (*MC*) samples are taken. To prevent wasted samples and simulations, MC sampling should stop when the CI *lower bound* ≥ target yield, as shown here, or when the *upper bound* ≤ target yield

One challenge in step 2 remains: when the optimizer considers a set of candidate spec values, it must calculate the overall yield. When there are ≫1 outputs, this is not trivial: density estimation scales poorly on >1 dimensions. There is a pragmatic solution, but the explanation is lengthy. So, we have inserted it into Appendix A of this chapter for the interested reader.

The overall outcome of the algorithm is a set of 3σ corners, with one corner for the min and/or max of each output. Each individual corner might have different partial yields, such as 3.1, 3.3, and 3.2σ; but together they lead to an overall yield of 3σ. The approach inherits the excellent speed, accuracy and scalability properties of single-output 3σ corner extraction. It is a key part of enabling the sigma-driven corners flow of 3σ corner extraction, rapid design iterations against the 3σ corners, and finally 3σ verification.

Another key part of the sigma-driven corners flow is 3σ verification, which we now describe.

4.4 Confidence-Driven 3σ Statistical Verification

Verification is a key step in the design flow, having the general aim to be satisfied that that circuit will work under the conditions of interest.

How might we think about verification in a statistical sense? Recall the challenge posed in Chap. 3: *Consider if we took* 10 *Monte Carlo (MC) samples, and all* 10 *were feasible. This gives a yield estimate of* 100 *%. Does that mean that we should trust the design to really have* 100 *% yield?*

As Chap. 3 described, the key is in confidence intervals (CIs). We can make a statistically sound estimate of the lower bound for yield, and of the upper bound for yield. For example, if 10/10 MC samples are feasible, then the upper bound is 100 %, but the lower bound is 72.2 % using the Wilson approach. If our yield target is 99.86 %, then we have not yet demonstrated a pass or a fail yet. But as we take more MC samples, then the confidence interval tightens.

With enough MC samples, the circuit will be verified to pass or fail,[7] in one of two ways:

1. If the lower bound is higher than the target yield, then the circuit passes
2. If the upper bound is lower than the target yield, then the circuit fails

A reasonable question is: how many MC samples are needed to verify the yield? There is no direct answer, because it depends on the circuit.[8] However, we can reframe the question towards the user task of verification. Specifically, the best response is to simply run MC sampling until one of the two conditions above is met, then stop. Figure 4.19 illustrates this concept. As the MC sampling tool runs and more MC samples are simulated, the width of the yield-estimate confidence interval tightens.

A competently designed MC sampling tool should support this type of verification behavior, rather than asking the user to monitor convergence. Also, the tool should support >1 environmental points and >1 test benches for each MC sample; a MC sample only passes if every output across all test benches passes spec on every environmental point. A competently-designed tool would also show the convergence over time, as in Fig. 4.19, automatically updated during the course of the MC run, to give the user intuition about how close the circuit is to passing or failing. Finally, the tool should make it easy to change the output specifications "on-the-fly" as the user learns about how well the circuit is yielding, and about the sensitivity to different specifications.

4.5 Optimal Spread Sampling for Faster Statistical Estimates

This section describes Optimal Spread Sampling (OSS), which reduces the average number of samples needed to estimate mean, standard deviation, and yield accurately.

4.5.1 Pseudo-Random Sampling and Yield Estimation

As Chap. 3 introduced, Monte Carlo (MC) circuit problems model process variations as statistical distributions. A typical MC run draws process points from the statistical distribution, then simulates those points using SPICE. Each simulated point returns a vector of outputs such as power consumption, gain, slew rate, etc. Using the returned output values from all process points, statistical estimates can

---

[7] Within a target confidence level, such as 95 %; which means that 19 times out of 20 the conclusion is valid.

[8] Though an approximate number, making light assumptions, is ≈ 80 MC samples for 2σ, and $\approx 1,400$ MC samples for 3σ. Section 4.6.1 provides a detailed answer.

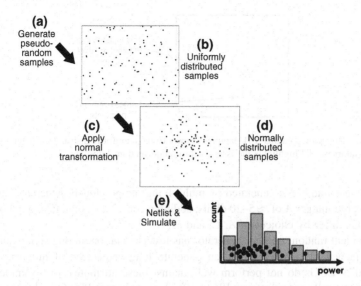

Fig. 4.20 The flow of pseudo-random sampling to estimate output yield, mean, etc. (**a**) Pseudo-random sampling from a 2-dimensional uniform distribution, to create (**b**) uniformly distributed samples. These samples are (**c**) transformed to be normally-distributed (**d**), e.g. via an the inverse cumulative distribution function (CDF) of the normal distribution. (**e**) For each sample, a net list is created by filling in the normally-distributed values into a template net list; and that net list is simulated. (**f**) From each sample and simulation, one or more output values are measured (e.g. power, gain, slew rate). All output values may be collected together to visualize the distribution, such as in a histogram. Furthermore, statistical estimates such as yield, mean, and standard deviation may be computed from the output values. Sensitivities of outputs to variables may be computed from the sample and output values

be made based on the outputs, such as average power, standard deviation of power, partial yield of power (percentage of samples that meet the power specification), and even estimates across all outputs at once, such as overall yield.

Typically, device-level process variation is modeled as a Gaussian distribution. Since most random number generators output uniformly-distributed values, they must be converted to a Gaussian distribution via a transformation, such as using the inverse CDF of the Gaussian, or the Box-Muller approach.

Figure 4.20 illustrates the overall flow, from generating uniform samples, converting to normal (Gaussian), simulating, and estimating the distribution of circuit performances, for the case of two random process variables.

Pseudo-random number generators (PRNGs) generate the uniform random numbers. The "pseudo" means the numbers are not truly random; they must be generated algorithmically to approach the ideal random. Linear congruental generators (LCGs)(e.g. Park and Miller 1988) are some of the oldest and best-known PRNGs, and have traditionally been how "rand()" functions are implemented. They work as follows. First, they start with a seed number, an integer. Then, each

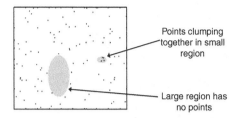

Fig. 4.21 This plot contains 101 samples drawn from a uniform distribution using pseudo-random samples, and highlights the issues with pseudo-random sampling

new integer number is generated by multiplying the previously-generated number with a large integer A of ≈ 5–20 digits, adding a constant C, then taking a modulus M. LCGs differ by choices of A, C, and M.

The ideal random number generator has low correlation among serial values in the sequence of numbers, and can generate long sequences of numbers (long "period"). LCGs do not perform well against these attributes, for example with periods typically ranging from 10^4 to 10^9. A more recent PRNG is the Mersenne Twister (MT) (Matsumoto and Nishimura 1998), which has low serial correlation and a vastly larger period of 10^{6000}. MT is becoming more widely adopted; for example, it has been in the Python programming language since version 2.3, when it replaced the Wichmann-Hill LCG.

4.5.2 Issues with Pseudo-Random Sampling

Figure 4.21 highlights the issues with pseudo-random sampling. It is easy to observe that the spread of the samples is very poor; some points cluster and nearly overlap, whereas other regions have no points at all. This leads to estimates of mean, standard deviation, and yield with higher estimation error than necessary.

For example, if making a yield estimate from 100 MC samples, the region with failures might be over-represented, leading to more MC samples with failures, and therefore a too-pessimistic estimate of yield. Or, it could under-represent failures, leading to a too-optimistic estimate of yield. As the number of samples increases, on average, each region will get represented more evenly; but it can take many samples. This means that the confidence interval for the yield will shrink fairly slowly, and therefore yield verification will take more simulations than needed.

4.5.3 The Promise of Well-Spread Samples

Monte Carlo sampling does not need to be random; it just needs to get a set (or sequence) of points that best cover the distribution it is sampling from. By

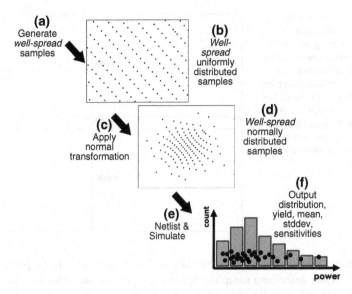

Fig. 4.22 *Well-spread* sampling replaces pseudo-random sampling (of Fig. 4.20) for better estimates (*lower variance*) of the output distribution, yield, mean, stddev, etc. This example uses Optimal Spread Sampling, which is described in a subsequent section

example, consider Fig. 4.22, which shows how a well-spread set of points may be generated, and how it translates to better estimates of yield, mean, etc.

Let us return to our yield example. Recall how poorly-spread samples lead to an over- or under-representation of the failure region, leading to a too-pessimistic or optimistic estimate of yield, and how that translated into confidence intervals that tightened slowly, taking more simulations than needed. If, on the other hand, we had well-spread samples, then the failure region would be well represented (not over- or under-represented), leading to a more realistic estimate of yield, leading to confidence intervals that tighten more quickly, and therefore taking fewer simulations than pseudo-random sampling.

4.5.4 The "Monte Carlo" Label

In our experience, the label "Monte Carlo" has taken two meanings in the circuit design and CAD field:

1. "Monte Carlo" applies to the general tool and approach to estimate yield, etc. by drawing samples from a distribution.
2. "Monte Carlo" is often also used to mean the pseudo-random sampling approach, as opposed to other approaches which might be more well-spread, such as Optimal Spread Sampling, discussed below.

Fig. 4.23 Four steps of *sequential* well-spread sampling, starting in the *top left* and going *clockwise. Note* how each subsequent step embeds the previous samples, and optimally fills the rest of the space with new samples. This example uses Optimal Spread Sampling, which is described in a subsequent section

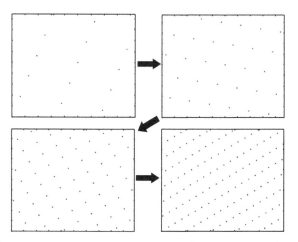

The latter label is not quite technically accurate, but seems to be pervasive in the field. When describing sampling approaches, we will use "Monte Carlo" and "pseudo-random" sampling interchangeably.

4.5.5   Well-Spread Sequences

In the course of performing a MC run using a well-spread sampling technique, we want the run to be well-spread not just when it hits some final target number of samples, but also during intermediate samples, so that representative estimates of yield and yield CIs can be made during the run. We want a well-spread *sequence*. Figure 4.23 shows a desirable behavior. Initial points in the sequence are broadly spread throughout the space, as shown in the top left. The second round of points, in the top right, doubles back to optimally fill in the gaps among the first round of points. The third round of points optimally fills in the gaps among the first and second points.

We aim to use a technique that can generate such well-spread point sets *and* sequences.

4.5.6 Low-Discrepancy Sampling: Introduction

In the literature, well-spread sampling is most commonly known as low-discrepancy sampling. The lower the discrepancy, the better that samples are spread. Sampling can be done to generate a single point set holding N items, or to generate a point sequence one sample at a time, and continuously make estimates using those samples. Common techniques include Latin hypercube sampling (LHS) (McKay et al. 1979; 2000), and quasi-Monte Carlo (QMC) (Sobol 1967).

The CAD community has explored application of these techniques to circuits (e.g. Singhee and Rutenbar 2010). However, LHS only gives payoff when the interaction among process variables is weak, and QMC scales poorly beyond 10 process variables.

Optimal Spread Sampling (OSS) is a low-discrepancy sampling technique that draws ideas from both digital nets and from lattice rules (L'Ecuyer and Lemieux 2000), drawing on advances from those fields rather than the older LHS and QMC approaches tried in the CAD community. The recent advances give it properties that are superior to older LHS and QMC techniques. OSS generates points with good spread in all the dimensions simultaneously rather than just one dimension at a time like LHS. It can scale to thousands or hundreds of thousands of input variables without resorting to heuristics like the recent QMC circuit techniques (Singhee and Rutenbar 2010, Veetil et al. 2011).

Appendix B gives a detailed review of low-discrepancy sampling, and details of OSS.

4.5.7 OSS Experiments: Speedup in Yield Estimation?

This section compares OSS to pseudo-random sampling and Latin Hypercube sampling (LHS), in the number of samples needed to estimate yield within 1 % accuracy.

The experimental setup included five representative circuits: GMC filter, comparator, folded opamp, current mirror, and low noise amplifier. Global and local variation were captured via the back-propagation of variance (BPV) model (Drennan and McAndrew 2003) on a modern industrial process. Each device had approximately 10 local process variables. Each circuit had reasonable width and length sizings. 1,500 pseudo-random (MC) samples were drawn and simulated with Synopsys® HSPICE® (Synopsys 2012). Then, specifications on outputs (e.g. gain, power) were picked such that yield was precisely 95 %. Then, a run of pseudo-random sampling was done, simulating and estimating yield one sample at a time. Being a Monte Carlo algorithm, as the sample size increases, the accuracy of the estimate improves on average. As soon as the run's estimate of yield was less than 1 % different than the reference yield of 95 %, the run was stopped and the number of simulations counted. This was repeated for a total of 30 runs of pseudo-random samples. Similarly, 30 runs of LHS were done, and 30 runs of OSS were done.

Table 4.1 shows experimental results comparing the average number of samples needed to reach an error within 1 % of the correct value. As we see, OSS has average speedups ranging from 1.19x (19 % faster) to 10.0× (900 % faster) compared to pseudo-random (MC) sampling. Recall that LHS does better when there is little interaction among process variables in the mapping to outputs; in our results it is the two amplifier circuits, which are known to have near-linear mappings. In contrast, LHS does considerably worse on the more nonlinear circuits,

Table 4.1 Summary of experimental results in estimating yield

| Circuit | # process variables | # samples to hit average of 1 % error | | | OSS speedup = #MC/#OSS |
		# MC samples	# LHS samples	# OSS samples	
GMC filter	1,468	285	215	65	4.38x
Comparator	639	325	255	180	1.81x
Folded opamp	558	295	250	245	1.20x
Current mirror	22	550	440	55	10.0x
Low noise amp	234	95	50	80	1.19x

such as the current mirror where it took 440 samples on average compared to OSS's 55.

4.5.8 OSS Experiments: Convergence of Statistical Estimates

This section compares Optimal Spread Sampling (OSS) to pseudo-random sampling by analyzing error convergence versus sample number for a variety of representative circuits.

The experimental setup for each circuit was as follows. We did 30 runs of OSS, where each run had 1,000 samples. We did 30 runs of pseudo-random sampling, where each run had 1,000 samples as well. Each run had a different random seed. We pooled together all samples from the 30 OSS runs and 30 pseudo-random runs, and from the pooled data, we measured mean, standard deviation, and yield.[9] We treated these measures as our "golden" reference values of mean, standard deviation, and yield.

On each run of either OSS or pseudo-random (MC), at a given number of samples, we estimated mean, standard deviation, and yield. The relative error was the estimated value, divided by its "golden" reference value. Then, the average error is taken across 30 runs. We computed average error for both OSS and pseudo-random, from 50 samples to 1,000 samples.

We plotted average error for OSS versus number of samples, as shown, for example, in Fig. 4.24's bottom solid line curve. On the same plot, we also plotted average error for pseudo-random versus number of samples as the top solid line curve. The plot is log–log so that the trends can be observed in linear form. To further facilitate comparison, we also performed a least-squares linear fit on the OSS curve, and the pseudo-random (MC) curve. These are shown as the dotted lines in the plot.

---

[9] In the case of yield, we actually chose a spec value such that the true yield value would be ≈ 95 %.

Fig. 4.24 Convergence of pseudo-random sampling versus OSS in estimating mean of bandwidth (*bw*) on a variable gain amplifier. OSS is the *lower curve*

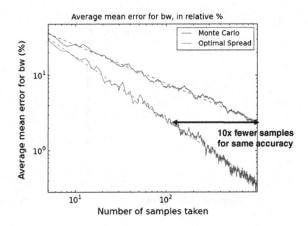

Figure 4.24 compares the convergence rate of OSS versus pseudo-random sampling. This figure shows convergence of estimating the mean value of a variable gain amplifier's bandwidth (VGA bw). From the convergence curves, we can estimate average speedup for a given accuracy. For example, if the target accuracy is 2 %, we find where the OSS dotted line intersects y = 2 %, which is 100 samples. The pseudo-random dotted line intersects y = 2 % at 1,000 samples. Therefore the speedup is 1,000/100 = 10x. In other words, OSS needed on average 10 × fewer samples than pseudo-random to estimate bandwidth within 2 % error.

We can also compare the accuracy achievable for a given simulation budget. For example, with x = 100 simulations, OSS gets an error of y = 2 %, whereas MC gets an error of 8 %. In other words, OSS had 4× lower error than pseudo-random for estimating bw at 100 simulations.

The slope for OSS is steeper than the slope for pseudo-random sampling. This means that the speedup of OSS over pseudo-random sampling gets exponentially faster as the number of samples increases. At 0.5 % error, OSS would need 800 samples and pseudo-random would need 50,000 samples, giving a speedup of 50,000/800 = 62x.

Figure 4.25 left compares the convergence of OSS and pseudo-random sampling on the VGA bw, in terms of estimating standard deviation. On the right, it is in terms of estimating yield. In both cases, OSS converges exponentially faster since its slope is steeper. Sometimes the curves jump around a bit, as in the left plot. This is simply statistical noise; the curves become more linear as more runs are done. Recall that we did 30 runs for each technique; with fewer runs the curves jump around more, and with more runs the curves jump around less.

Figure 4.26 compares OSS and pseudo-random sampling on two other problems—gain of the VGA, and delay of a ring oscillator. Once again, OSS outperforms pseudo-random sampling. We have benchmarked a wide variety of analog, custom digital, and memory circuits, and OSS always outperforms pseudo-random sampling.

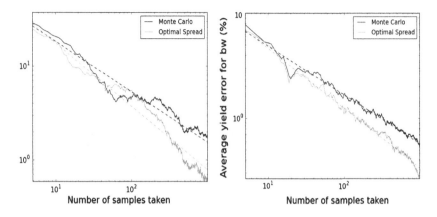

Fig. 4.25 Convergence of pseudo-random sampling versus OSS in estimating standard deviation of bw on a VGA (*left*), and partial yield of bw on a VGA (*right*). In both subplots, OSS is the *lower curve*

4.5.9 Assumptions and Limitations of Optimal Spread Sampling

To our knowledge, OSS makes no assumptions about the nature of the sampling problem, and has no obvious theoretical limitations. One practical limitation exists, which is that the amount of speed and/or accuracy improvement from using OSS varies from circuit to circuit, and cannot be predicted in advance. However, as the Experiments section demonstrated, OSS is superior to both pseudo-random and Latin Hypercube methods in almost all cases. Even in the rare case where OSS is not the superior sampling technique, OSS will still converge on the correct yield estimate, just more slowly. This makes OSS a very safe, low-risk, high-reward technique to employ.

In terms of circuit characteristics, there are no limitations with OSS. OSS works on problems with a small or large number of samples. OSS also works on a small or large number of variables; this is in contrast to most QMC approaches (e.g. Sobol) which only have good spread for the first 10–15 dimensions. OSS works on linear, weakly nonlinear, or strongly nonlinear circuits with significant variable interactions; this is in contrast to Latin Hypercube sampling which does not explicitly account for variable interactions. This also means that Latin Hypercube is most competitive to OSS on problems with little variable interaction. If there was a degenerate case, then OSS performance would simply reduce to that of pseudo-random.

In terms of performance, the runtime cost for generating a set or sequence of OSS samples is about the same as drawing pseudo-random samples. OSS takes some up-front computational time to compute z (how to space each dimension). After that, all the other computations are near-trivial. This time is not noticeable from the user perspective, even for fast-simulating circuits. Pseudo-random sampling has no up-front computation, but generating takes slightly more time than

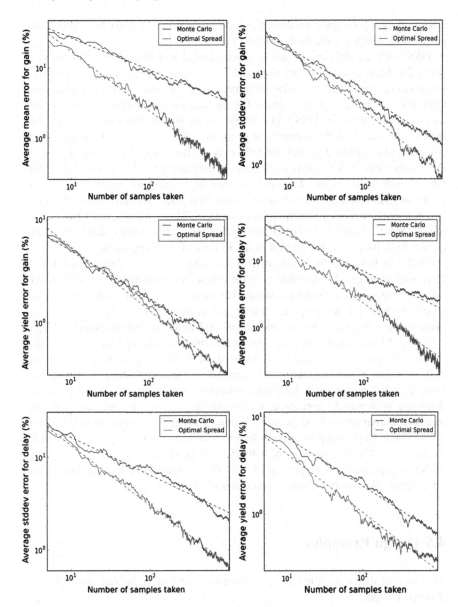

Fig. 4.26 Convergence of pseudo-random sampling versus OSS in estimating (**a**) mean of gain on a VGA (**b**) std. dev. of gain on a VGA (**c**) partial yield of gain on a VGA (**d**) mean of delay on a ring oscillator (**e**) std. dev. of delay on a ring oscillator (**f**) partial yield of delay on a ring oscillator. In all subplots, OSS is the lower curve

OSS in drawing samples because d pseudo-random values must be drawn, compared to basically d calls to mod() for OSS.

OSS has been shipping as part of a commercial tool since 2010, and has been applied to hundreds or possibly thousands of different circuits. OSS comes with a usage caveat: one cannot *conveniently* measure its speedup. Upon learning about OSS, users sometimes want to know the speedup on *their* circuit. To do so, they may do a single run of OSS and a single run of pseudo-random sampling, then eyeball the value of the estimate (e.g. mean) over a number of samples, then sometimes be surprised to find that OSS may not have converged quite as quickly as pseudo-random. Since it is a stochastic algorithm, a single run tells very little. One should never expect to conclude anything significant about OSS, Monte Carlo, or LHS convergence with a single run from each. We have found that there is a fair amount of run-to-run variance in OSS, Monte Carlo, and LHS in estimating mean, standard deviation, and yield, and to benchmark them with sufficiently high precision, we needed to do 30 runs of each approach.

OSS does have benefit, but the benefit is averaged over many runs. Therefore, if there is on average a $5\times$ speedup, over the course of many circuit designs there will be an average of $5\times$ reduction in simulation usage for the same quality. But, it will simply be hard for designers to reliably and qualitatively see the benefit on any given run. For these reasons, we recommend OSS as the default sampling technique.

Doing 30 benchmarks of each approach is time-consuming and tedious, and at the end, designers doing single runs will still not be able to qualitatively report that OSS is better. If one is looking to benchmark the speed of Monte Carlo approaches, then we recommend 3σ corner extraction for far greater speed gains. By designing with true 3σ corners, at each candidate design, one only needs to simulate one corner for each output (say, 5 total). That can be compared to simulating 1,400 Monte Carlo samples at each design candidate; 1,400 samples is enough for 3σ accuracy. This is a speedup of $1,400/5 = 280x$. Even if one was comparing to 50 MC samples, that is a speedup of 10x. The designer benefit is qualitative and tangible: it allows rapid design iterations with 3σ accuracy.

4.6 Design Examples

This section discusses three design scenarios based on industrial application of 3-sigma design.

4.6.1 How Many Monte Carlo Samples?

When running a Monte Carlo analysis, how many samples are enough? 30? 100? 1,000? A common default is to use 100 samples; however, most of the time 100 samples is either too many or not enough.

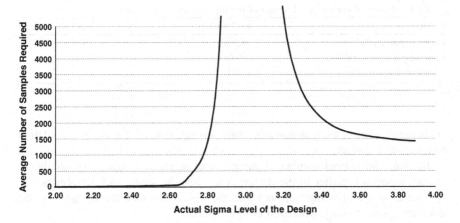

Fig. 4.27 Average number of samples required to verify to 3 sigma

The correct number of samples to use depends on several factors, including the design goal (e.g. design verification, statistical corner extraction, quick smoke test) and the yield target for the design.

For example, consider the following scenario: A design is to be verified with Monte Carlo to determine if it is robust to 3 sigma (approx. 99.7 % yield), within a 95 % statistical confidence. The minimum number of samples required to do this successfully depends on the actual yield of the design. If the design is well above 3 sigma, then a minimum of $\approx 1{,}400$ samples is needed in order to conclude with sufficient confidence that the design meets its 3-sigma target using binomial pass/ fail statistics. However, if the design is well below 3 sigma, then only a very small number of samples is needed; approximately 20 or so, enough to demonstrate that there is no chance of ever reaching 3 sigma. If the design is very close to 3 sigma, then the required number of samples increases significantly; thousands or even tens of thousands of samples may be required. Figure 4.27 illustrates.

It is important to note from this example that the correct number of samples cannot be determined *a priori*, since the correct number of samples is dependent on the true yield of the circuit, which is not known until the Monte Carlo analysis is started. Therefore, it is necessary to determine the correct number of samples on-the-fly while Monte Carlo analysis is running.

In practice, the correct number of samples can be enforced either by: (1) determining the relationship between the minimum number of samples and yield for the desired sigma target, and then manually monitoring the simulation results as they complete; or by (2) using design software that automatically determines and simulates the correct number of samples on-the-fly and stops when exactly the minimum required number of samples has been run. If a limited amount of time or resources is available for simulation, an upper bound on the number of simulations can be used in combination with one of these two methods. This approach provides

Table 4.2 Result of corner-based design of the flip-flop for each method

Method	# of simulations	Final sigma level
Direct monte carlo	16,800	>3
Lightweight monte carlo	2,500	2.73
Sigma-driven corners	1,707	>3

as thorough a verification as possible while staying within the available simulation time/resource budget.

Running the correct number of samples can result in a significant simulation saving, especially earlier in the design flow when the sizing and performance of the design are still being adjusted and the sigma level of the design may be fluctuating significantly with each design iteration.

4.6.2 Corner-Based Design of a Flip-Flop

In this example, using a flip-flop design, the circuit topology has been set and initial transistor sizes selected, and it is now necessary to design for statistical variation. Three different methods for doing this are compared. To facilitate a reasonable comparison, the number of design iterations for each approach is limited to no more than ten.

Method 1: Direct Monte Carlo. In this method, a direct Monte Carlo analysis is used for both verification and design. As discussed earlier, to verify accurately to 3 sigma requires a minimum of approximately 1,400 samples. Therefore, to have high certainty that the design is meeting the 3-sigma target, 1,400 samples need to be run for each design iteration. Despite the accuracy of the result, this approach is highly inefficient and impractical.

Method 2: Lightweight Monte Carlo. For this method, design verification is done with a thorough Monte Carlo analysis, but design iterations are performed using fewer Monte Carlo samples (100 samples per run). After ten design iterations, verification shows that the final sigma level of the design falls short of the 3-sigma target, meaning that either additional iterations are required, or the design will not be sufficiently robust.

Method 3: Sigma-driven corners. In this method, design iteration is performed using corners obtained from sigma-driven corner extraction. This results in fewer simulations than the other methods, and provides a final design that meets the 3-sigma design target.

Table 4.2 summarizes the number of simulations and final sigma level of the flip-flop design for each method. Of the three methods, the sigma-driven corners flow is the only one that is simultaneously fast (low number of simulations), and accurate (hit 3 sigma).

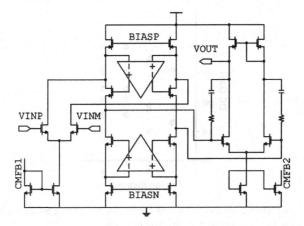

Fig. 4.28 Folded-Cascode amplifier with gain boosting

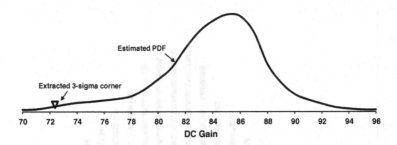

Fig. 4.29 3-sigma extracted corner for DC gain. The estimated probability density function (PDF) for DC gain is also shown

4.6.3 3-Sigma Design of a Folded-Cascode Amplifier

In this example, we revisit the folded-cascade amplifier design that was discussed in Chap. 2. The schematic is shown again in Fig. 4.28 for convenience.

The goal for this design is now to find a good tradeoff among performances at 3 sigma. In particular, it is desirable to balance gain, phase margin, and power consumption without adversely impacting other characteristics of the amplifier (e.g. bandwidth, noise rejection).

First, sigma-driven corner extraction is used to find 3-sigma corners for the design. This determines statistical corners for each output subject to the overall sigma target of 3 sigma for this design. Figure 4.29 shows the 3-sigma extracted corner for DC gain, along with the estimated probability density function (PDF) obtained using sigma-driven corner extraction. Note that the DC gain distribution is non-Gaussian, and that the extracted 3-sigma corner accounts for this.

Sigma-driven corner extraction for this design needs only 47 simulations to obtain a representative set of five 3-sigma corners corresponding to the five key outputs of interest.

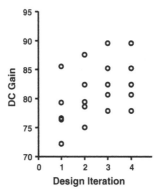

Fig. 4.30 DC gain performance across 3-sigma corners for four design iterations. Each point in the plot is the result of simulating one of the 3-sigma corners

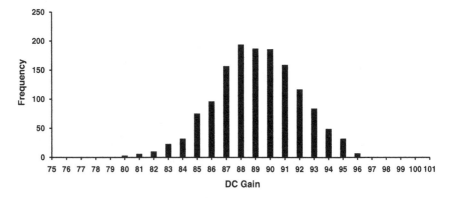

Fig. 4.31 Final DC gain distribution

Next, the extracted corners are used to identify sensitive devices in the design. Sweeping design parameters and simulating them across the extracted corners reveals which design parameters (i.e. w, l, r, c, etc.) lead to the best tradeoffs between design performances at 3-sigma. For this design, to sweep across key devices, across 3-sigma corners, with five sweep steps per parameter, requires only 245 simulations. This is significantly fewer simulations than would be required to perform a 3-sigma Monte Carlo analysis for each sweep point, yet it provides effectively the same information.

Design iterations are then performed based on the sweep results. Each time a change is made to the design, the five 3-sigma corners are re-simulated. This makes it possible to observe 3-sigma tradeoffs while making changes to the design, so that a proper balance between performances, power consumption, and area can be achieved. Figure 4.30 shows the DC gain performance across the five 3-sigma corners for four design iterations. Improvement can be seen with each iteration except for the last one. Performing these four design iterations requires only 20 simulations (5 corners × 4 iterations).

Once the design iterations are complete, statistical verification is performed to confirm the final design performance to 3 sigma. Figure 4.31 shows the final DC gain distribution from 1,420 Monte Carlo samples, which reflects the improvement achieved by iterating with the 3-sigma corners. In this case, no additional iterations are necessary.

3-sigma DC gain is improved by 8 %. Equally important, this improvement is achieved while maintaining satisfactory performance for other design characteristics, including phase margin, bandwidth, and power consumption, under 3-sigma conditions. The entire process of sigma-driven corner extraction, sensitivity analysis with the extracted corners, and design iteration, takes only 312 simulations, followed by the single verification run with 1,420 simulations. This makes for a very efficient yet accurate 3-sigma design flow for this design.

4.7 Conclusion

This chapter reviewed various design flows to handle 3σ statistical variation and described how the sigma-driven corners flow gives the best combination of speed and accuracy. It then described the components that enable the sigma-driven corners flow: sigma-driven corner extraction and confidence-driven 3σ verification. It then described Optimal Spread Sampling (OSS), which speeds up the accurate estimation of mean, standard deviation, and yield. Finally, it presented two design examples.

Appendix A describes density-based yield estimation on >1 output, which enables 3σ corner extraction on >1 output. Appendix B describes details of low-discrepancy sampling and OSS.

Appendix A: Density-Based Yield Estimation on >1 Outputs

Introduction

This section discusses the challenge of yield estimation, with a focus on the application to corner extraction (Sect. 4.3.3). As we will see, density-based approaches provide the necessary resolution, but need special consideration for >1 output. Of the possible approaches to handle >1 output, the "Blocking Min" approach provides the requisite speed, accuracy, and scalability.

Given a set of Monte Carlo (MC) sample output values, there are two main ways to estimate yield: binomial and density-based. Chapter 3 introduced these

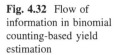

Fig. 4.32 Flow of information in binomial counting-based yield estimation

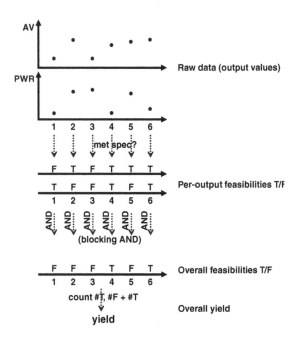

approaches, and included a description of how confidence intervals for each approach were calculated. It did not discuss how a corner extraction algorithm might use yield estimates, or how density estimation might handle >1 outputs. This section covers those topics.

Binomial-Based Yield Estimation on >1 Outputs

In the binomial approach to estimate yield, one counts the number of MC samples that are feasible on all outputs, and the total number of samples. The yield estimate is simply the ratio of (number feasible)/(total number).

From this simple description, Fig. 4.32 shows information flow in a more detailed fashion. We will be using this view as a framework to present a new technique for yield estimation. At the top of Fig. 4.32, we have 6 Monte Carlo samples, each which has a value for output AV and for output PWR. Each Monte Carlo value for AV and for PWR is compared to its spec, and marked as feasible = True (T), or feasible = False (F). Then, on each sample, the T/F value for output AV is merged with the T/F value for PWR, via the AND operator (only T if both input values are T). This is actually a *blocking* operation in the statistical sense, because the blocks of data that are similar to one another—the T/F value for each output within each MC sample—are kept together. Now we have one T/F value for each Monte Carlo sample. The yield estimate becomes the ratio of (number feasible, or T's)/(total number, or T's and F's).

Fig. 4.33 Computing yield
from a one-dimensional
density-estimated PDF

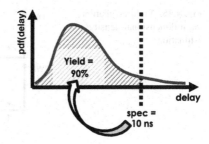

Fig. 4.34 Flow of
information in density-based
yield estimation: the
challenge

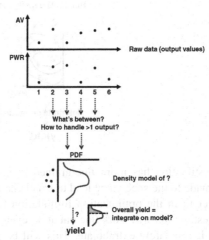

Recall that we are estimating yield, with an eye towards the application of 3σ
corner extraction. In 3σ corner extraction, we want to be able to make small
changes to specifications out at $\approx 3\sigma$, and get back slightly different estimates for
yield. That is, it needs fine-grained precision at 3σ. This is especially necessary in
an optimization formulation for corner extraction (Sect. 4.4.3). The problem is that
for the binomial approach to start to have good precision out at 3σ, it needs 1,000
samples or so. While 1,000–2,000 samples are reasonable for verification, that is
quite an expensive demand for corner extraction, which does not need to be *as*
accurate as verification.

Since a binomial MC approach does not provide us with the desired resolution for
3σ corner extraction, let us examine density estimation to see how well it might fit.

Density-Based Yield Estimation on One Output

We first discuss the density-estimation approach to estimate yield on one output,
then consider how we might handle >1 outputs. Figure 4.33 reviews how yield is
calculated from a density-estimated PDF. Quite simply, it is the result of integrating under the PDF in the range from $-\infty$ to the spec value (or spec value to

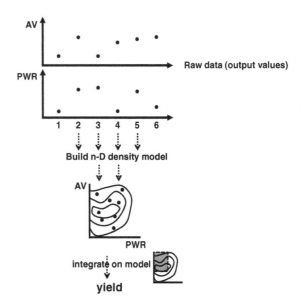

Fig. 4.35 Yield estimation via n-dimensional density estimation

+∞). Density estimation has more fine-grained precision at 3σ than binomial: small changes made to the spec value lead to small changes in yield estimate. This makes it a better fit in the optimization formulation for corner extraction (Sect. 4.4.3). Density estimation will work fine out at 3σ even with 50–100 MC samples; it assumes that it can safely extrapolate. This will be accurate on some circuits, where performance does not drop off sharply. On other circuits it will not be as accurate, but for corner extraction that's fine because the verification step will catch it. (And if there is failure in the verification step, then the new corner extraction round will have better accuracy because it will have more MC samples to work from.)

Density-Based Yield Estimation on >1 Output

When we consider using density estimation to compute a yield across >1 outputs, the discussion gets more complex. Figure 4.34 illustrates the target flow. At the top, we have raw MC output values coming in. Out the bottom, we want to build *some* sort of density model or models, and somehow integrate on that, to get an overall yield estimate. The question is, what are the appropriate steps in between. It turns out there are a few different options to accomplish this, with different pros and cons. Let us examine the options.

One approach is to do *n*-dimensional density estimation across the *n* outputs, then simply integrate directly. Figure 4.35 illustrates. The challenge with this approach is that density estimation has poor scalability properties: the quality of the density models degrades badly as dimensionality increases. Even 5 dimensions

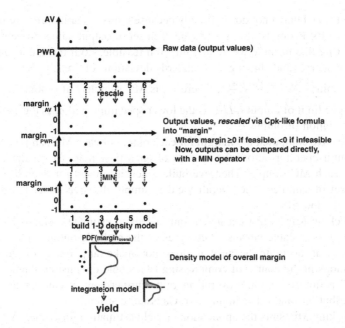

Fig. 4.36 Flow of information in density-based yield estimation: the solution is blocking min

gives quite poor density models, and many circuit applications might even have ≫5 outputs.

Another approach is to estimate the PDF of each output one at a time, then combine them somehow. This is a subproblem in some approaches to do statistical static timing analysis (SSTA), which need to combine two input delay PDFs into a single "worst-case" (max) delay PDF. The idea is to approximate the "max" operator with a linear function, where the linear function is calibrated by the incoming PDFs. This approach has been extended to analog circuits, and the "Linear Max" function extended to a "Quadratic Max" (Li and Pileggi 2008). However, this approach only handles unimodal PDFs, and the "max" operator induces error which degrades the overall accuracy of yield estimation.

The final approach is a novel technique which we call the "Blocking Min". It does not suffer from the accuracy issues of the other approaches, and scales to an arbitrarily high number of outputs. The core idea is as follows. Looking back to the Linear/Quadratic Max approaches, we see that PDFs are estimated one at a time, then combined. This ignores the natural groupings of outputs into individual MC samples. In contrast, the Blocking Min exploits these natural groupings. The general idea is to keep the natural groupings together and apply the "min" operator to real MC sample values, to compress >1 outputs into a single scalar "combined" output. Only at this point is a PDF estimated, from "combined" output.

Figure 4.36 illustrates the Blocking Min approach. At the top is the raw data, with a value of AV and PWR for each MC sample. We want to apply a "min" operator to these, but cannot do so directly because the AV's units (dB) are not the same units as PWR (amps). So, we rescale each set of output values to be *margin* values in a *Cpk*-like formula where margin ≥ 0 if feasible, <0 if infeasible, and 0 if on the boundary, and having a standard deviation of ≈ 1.0. Specifically, $margin_{i,j} = \min\left(\frac{USL_i - v_{i,j}}{\hat{\sigma}_l}, \frac{v_{i,j} - LSL_i}{\hat{\sigma}_l}\right)$, where $v_{i,j}$ is MC sample j of output i, USL_i is the upper spec limit of output i, LSL_i is the lower spec limit, and $\hat{\sigma}_l$ is the estimated standard deviation of output i.

Once we have computed the margin for each output of each MC sample, we can apply the min operator across the outputs of an MC sample, to get the *overall* margin for each MC sample. Then, we build a density model for overall margin, from that set of samples. The overall yield is simply the area under the density model for a value ≥ 0.

The Blocking Min is fast because it only needs to estimate a single 1-dimensional PDF. It is scalable because it compresses the multiple outputs into a single dimension. It is accurate because it does not make any linear or quadratic approximations in the course of compressing to a single dimension, thanks to the "blocking" action to compute overall margin. Furthermore, it can handle multimodal distributions and other highly non-Gaussian distributions.

The Blocking Min suits the application of yield estimation for corner extraction quite naturally. It is fast, accurate, and scalable as discussed; and because it uses density estimation it provides high resolution to support an optimization-style tuning of specifications.

Appendix B: Details of Low-Discrepancy Sampling

This section has three parts: a detailed literature review of low-discrepancy sampling (Sect. B.1), followed by descriptions of Optimal Spread Sampling (OSS) for point *sets* (Sect. B.2) and for point *sequences* (Sect. B.3).

B.1: Detailed Review of Low-Discrepancy Sampling

In the literature, "well-spread" sampling is most commonly known as "low-discrepancy sampling". The lower the discrepancy, the better that samples are spread. A simple example of a discrepancy measure is the (negative) minimum distance between all points in a sample set; and more complex measures exist in the literature. Sampling can be done to generate a single point set holding N items, or to generate a point sequence one sample at a time, and make continuous estimates using those samples.

Low-discrepancy sampling has origins from Quasi Monte Carlo (QMC) methods as well as cubature methods, which were both developed with a focus on numerical integration. Modern low-discrepancy techniques can be classified into two main categories: Digital Nets and Lattice Rules. Digital Nets encompass many traditional QMC techniques, including Halton (1960), Sobol (1967), Faure (1982), Hammersley (1960), Niederreiter (1987), and Niederreiter-Xing sampling (1998). Lattice Rules encompasses many traditional techniques including orthogonal arrays, Latin Hypercube Sampling (McKay et al. 1979; 2000), and Latin Super-cube Sampling (Owen 1998). While Digital Net methods were traditionally designed for point sequences, they can be used for point sets; and while Lattice Rules methods were traditionally designed for point sets, researchers have shown how to alter them for use as point sequences (Cools et al. 2006).

The CAD field has explored some low-discrepancy sampling approaches. Latin Hypercube Sampling (LHS) (McKay et al. 2000) is quite simple and quite popular (Keramat and Kielbasa 1997; Jaffari 2011; Tao 2011; Liu and Gielen 2012). It works as follows. If one aims to generate n random samples in d dimensions, then each dimension gets divided into n bins of equal probability. When drawing the samples, each bin will get drawn from exactly once. Overall, this means that there will be good spread for each dimension, independently of other dimensions. However, LHS does nothing to ensure that there is good spread among *points* in >1 dimension. This matters when there are interactions among random (process) variables in the mapping to output variables. As we saw in Sect 4.6.7, LHS does well on circuits where the interactions are weak, and not as well on circuits with stronger or higher-order interactions. There are some techniques to improve LHS on second-order interactions (e.g. Jaffari 2011), but these increase complexity, increase runtime, and still do not handle higher-order interactions.

Other circuit CAD researchers have explored variants of QMC methods like Sobol' sampling. Many QMC methods do poorly in >10 or so dimensions, so the research has focused on workarounds to handle hundreds or thousands of dimensions. (Singhee and Rutenbar 2010) bypassed the issue by doing a short "pilot" run first to estimate the relative importance of each process variable, then focused the QMC sampling on the most important 10 dimensions. (Veetil et al. 2011) was similar, focusing QMC methods to the most important variables. Of course, this only helps if 10 process variables have most of the impact. (McConaghy 2009) showed a representative circuit problem where the first 10 variables only had 50 % of the impact, and it took 85 variables to get 95 % of the impact. A further challenge is that most circuits have >1 output. With 5 outputs, each with different high-impact variables, one would need $5 \times$ more "important" variables, or to assign just $10/5 = 2$ important variables per output.

Optimal Spread Sampling (OSS) is a low-discrepancy sampling technique that draws ideas from both Digital Nets and from Lattice Rules, drawing on advances from those fields rather than the older LHS and QMC approaches that other circuit references used. The recent advances give it properties that greatly improve older LHS and QMC techniques. It generates points with good spread in all the dimensions simultaneously, rather than just one dimension at a time like LHS. It can scale to thousands or hundreds of thousands of input variables, without resorting to heuristics like the recent QMC circuit techniques.

Table 4.3 Procedure UniformOssSet()

Input: number of points n, dimension d
Output: point set P (order does not matter)
1. z = minimize worst-case error on all functions
2. $u \sim \text{unif}^d(0,1)$
3. z = randomly permute z
4. $P = \emptyset$
5. for $k = 0, 1, ..., n-1$ // for each sample
6.   for $j = 1, 2, ..., d$ // for each dimension
7. $P_{k,j} = \{k*z_j/n + u_j\}$
8. return P

B.2: Creating a Point Set with Optimal Spread Sampling

This section describes how to create a set of uniformly-distributed n samples in d-dimensional space using Optimal Spread Sampling (OSS).

OSS gives the point set

$$P = \{k * z/n\}; \; k = 0, 1, ..., n-1$$

where $\{x\}$ is the fractional part of x, i.e. $\{x\} = x - floor(x)$; and $z = (z_1, ..., z_d)$ is the *generating vector*, a d-dimensional integer vector having no factor in common with n.

Once a z is determined, generating a point set P is straightforward, using the equation above. The challenge is to determine z for the given n and d. The OSS algorithm implicitly holds a Fourier-series approximation of all possible functions, and optimizes across all possible z to minimize the discrepancy measure of worst-case error (Sloan 1994). This optimization has time complexity $O(d\, n \log(n))$ and memory complexity $O(n)$, where d is the number of process variables and n is the number of samples.

Like many low-discrepancy approaches, OSS samples can be "randomized" so that it becomes a variance reduction technique taking samples in the uniform $(0,1)$ space, yet retains its high uniformity when taken as a point set. A simple way to do this is with the *random shift modulo* technique (aka Cranley-Patterson rotation) (Cranley and Patterson 1976), which draws a single d-dimensional point $u \sim unif^d(0,1)$ and adds it to the point set P, modulo 1. This operation can equivalently be incorporated into equation for the point set for the computation of P. A second randomization, permuting z before generating a new point set, will ensure that all variables are treated equally.

Table 4.3 presents the pseudocode to generate to generate a (randomized) set of n points P in $\text{unif}^d(0,1)$ space.

The vector z is computed in step 1 of Table 4.3. There are many techniques to accomplish this, from very early, simple techniques (Korobov 1959) to more involved modern, complex, but scalable (Sinescu and L'Ecuyer 2011). Step 2 and

Table 4.4 Procedure UniformOssSequence()

Input: number of points n, dimension d, base b (e.g. 2)
Output: point sequence P (order matters)
1. z = minimize worst-case error (embedded)
2. $m_1 = 1$
3. $u \sim \text{unif}^d(0,1)$
4. z = randomly permute z
5. $P = \emptyset$
6. $i = 0$
7. for $m = m_1, ..., m_2$
8.   for k = randomly permute $(0, 1, ..., b^m-1)$
9.   if $\text{mod}(k, b) \neq 0$ //point is new
10. $i = i + 1$
11.   for $j = 1, 2, ..., d$
12. $P_{i,j} = \{k^*z_j/b^m + u_j\}$
13. return P

step 7 accomplish Cranley-Patterson rotation for randomization; step 2 draws a point from a uniform distribution using pseudo-random sampling, and step 7 performs the actual rotation via the addition of u_j and the modulo operator $\{\}$. Step 3 ensures all variables are treated equally. Steps 4–8 iteratively build up the point set P. Each entry in P is a value for one variable d of one sample k. The key operation is step 7, where the base value is a multiple of k and z_j, down-sampled by the number of samples n (then "randomized" via the Cranley-Patterson rotation).

In our experience, the extra computational cost of this algorithm is negligible compared to the cost of pseudo-random number generation.

B.3: Creating a Point Sequence with Optimal Spread Sampling

Figure 4.23 introduced the possibility that Optimal Spread Sampling (OSS) could be used to generate not just *sets*, but *sequences* too. A sequence is desirable for "anytime" style algorithms, where each additional step of the algorithm provides incremental value to the user, rather than relying on the algorithm to complete fully before results are available. For Monte Carlo sampling, a sequence of well-spread points is useful to give on-the-fly information to the user during sampling, rather than waiting until a full sampling run is complete. This section describes how OSS sequences can be generated.

OSS sequences are possible when the upper limit on n can be estimated; this occurs in many practical problems such as when the user has pre-specified the number of process points, or the target yield to verify a design (from which the number of points can be estimated under mild assumptions). Then the core idea is to embed smaller point sets in successively larger point sets, as Fig. 4.23 shows.

In mathematical terms, if P_m is the point set from OSS with b^m points, then P_1 is a subset of P_2, is a subset of P_3, etc.

To use this practically, we choose a base value b (e.g. 2), let $m_1 = 1$, then compute minimal m_2 such that $b^{m_2} \geq n$. Then, we compute a z which works across a range of m values $m = \{m_1, m_1 + 1, ..., m^2\}$, to account for many possible point sets simultaneously. It has runtime $O(dn(log(n))^2)$. With z in hand, points are first drawn from set P_1, then set P_2/P_1, then set $P_3/(P_2 \cup P_1)$, and so forth. Each set's points can be ordered randomly, with gray codes, or with radical inverse (Niederreiter 1987).

Table 4.4 presents the pseudocode to generate a *sequence* of n points. Compared to the approach for sets (Table 4.3), it has an outer loop on exponent m (step 7), and each sample divides by b^m rather than by n (step 12). It uses i for the sample index (steps 6, 10, 12). Since each set P_{m+1} embeds all smaller subsets, it must avoid those; this turns out to be easy because the points in P_{m+1} whose indices k are multiples of b (step 9).

The Optimal Spread Sampling option in Solido Variation Designer draws samples with an OSS sequence.

References

Boggs PT, Tolle JW (1995) Sequential quadratic programming. Acta Numer, pp 1–50

Cools R, Kuo FY, Nuyens D (2006) Constructing embedded lattice rules for multivariate integration. SIAM J Sci Comput 28(6):2162–2188

Cortes C, Vapnik VN (1995) Support-vector networks. Mach Learn, 20

Cranley R, Patterson T (1976) Randomization of number theoretic methods for multiple integration. SIAM J Numer Anal 13(6):904–914

Cressie N (1989) Geostatistics. Am Stat 43:192–202

Drennan PG, McAndrew CC (2003) Understanding MOSFET mismatch for analog design. IEEE J Solid State Circuits (JSSC) 38(3):450–456

Faure H (1982) Discrepance des suites associees a un system de numeration en dimensions. Acta Arithmetica 61:337–351

Graeb H (2007) Analog design centering and sizing, Springer, Dordrecht

Halton J (1960) On the efficiency of certain quasi-random sequences of points in evaluating multi-dimensional integrals. Numer Math 2:84–90

Hammersley J (1960) Monte Carlo methods for solving multivariate problems. Ann NY Acad Sci 86:844–874

Hershenson MDM, Boyd SP, Lee TH (1998) GPCAD: a tool for CMOS op-amp synthesis. In: Proceedings of international conference on computer-aided design (ICCAD), pp 296–303

Jaffari J (2011) On efficient LHS-based yield analysis of analog circuits. IEEE Trans Comput Aided Des Integr Circuits Syst 30(1):159–163

Keramat M, Kielbasa R (1997) Latin hypercube sampling monte carlo estimation of average quality index for integrated circuits. Analog Integr Circ Sig Process 14(1–2):131–142

Korobov NM (1959) The approximate computation of multiple integrals. Dokl Akad Nauk SSSR 124:1207–1210 (In Russian; referenced by L'Ecuyer and Lemieux 2000)

L'Ecuyer P, Lemieux C (2000) Variance reduction via lattice rules. J Manage Sci 46(9): 1214–1235

Liu B and Gielen G (2012) A fast analog circuit yield estimation method for medium and high dimensional problems. In: Proceedings of design automation and test in Europe (DATE), Dresden, March 2012

Li X, McAndrew CC, Wu W, Chaudry S, Victory J, Gildenblat G (2010) Statistical modeling with the PSP MOSFET model. IEEE Trans Comput Aided Des Integr Circuits Syst 29(4):599–606

Li X, Pileggi L (2008) Quadratic statistical max approximation for parametric yield estimation of analog/RF integrated circuits. IEEE Trans Comput Aided Des Integr Circuits Syst 27(5):831–843, May 2008

Magma Design Automation (2012) Titan ADX Product Page, http://www.magma-da.com/products-solutions/analogmixed/titanADX.aspx (last accessed May 21, 1012). Magma is now part of Synopsys, Inc

Matsumoto M, Nishimura T (1998) Mersenne twister: a 623-dimensionally equidistributed uniform pseudo-random number generator. ACM Trans Model Comput Simul 8(1):3–30

McAndrew CC, Stevanovic I, Li X, Gildenblat G (2010) Extensions to backward propagation of variance for statistical modeling. IEEE Des Test Comput 27(2):36–43

McConaghy T (2009) Latent variable symbolic regression for high-dimensional inputs. In: Riolo R, O'Reilly U-M, McConaghy T (eds) Genetic programming theory and practice VII, Springer, NY (invited paper)

McConaghy T (2011) High-dimensional statistical modeling and analysis of custom integrated circuits. In: Proceedings of custom integrated circuits conference (CICC)

McKay M, Beckman R, Conover W (1979; 2000) A comparison of three methods for selecting values of input variables in the analysis of output from a computer code. Technometrics 42(1):55–61

Niederreiter H (1987) Point sets and sequences with small discrepancy. Monatshefte fur Mathematik, pp 104–133

Niederreiter H, Xing C (1998) The algebraic-geometry approach to low-discrepancy sequences volume 127, pp 139–160, Springer-Verlag, Berlin

Owen A (1998) Latin supercube sampling for very high-dimensional simulations. ACM Trans Model Comput Simul 8(1):71–102

Park SK, Miller KW (1988) Random number generators: good ones are hard to find. Commun ACM 31 10:1192–1201

Schenkel F et al (2001) Mismatch analysis and direct yield optimization by spec-wise linearization and feasibility-guided search. In: Proceedings of design automation conference (DAC), pp 858–863

Sinescu V, L'Ecuyer P (2011) Existence and construction of shifted lattice rules with an arbitrary number of points and bounded worst-case error for general weights. J Complexity 27(5):449–465

Singhee A, Rutenbar RA (2010) Why quasi-Monte Carlo is better than Monte Carlo or Latin hypercube sampling for statistical circuit analysis. IEEE Trans Comput Aided Des Integr Circuits Syst 29(11):1763–1776

Silva LG, Silveira LM, Phillips JR (2007) Efficient computation of the worst-delay corner. In: Proceedings of design automation and test in Europe (DATE), March 2007

Sloan IH (1994) Lattice methods for multiple integration. Oxford University Press, Oxford

Sobol I (1967) On the distribution of points in a cube and the approximate evaluation of integrals. Comput Math Math Phys 7:86–112

Synopsys Inc. (2012) Synopsys® HSPICE®, http://www.synopsys.com

Tao L (2011) A numerical integration-based yield estimation method for integrated circuits. J Semiconductors, vol 32

Veetil V, Chopra K, Blaauw D, Sylvester D (2011) Fast statistical static timing analysis using smart Monte Carlo techniques. IEEE Trans Comput Aided Des Integr Circuits Syst 30(6):852–865

Yao P et al (2012) Understanding and designing for variation in GLOBALFOUNDRIES 28 nm technology. In: Proceedings of design automation conference (DAC), San Francisco, June 2012

Zhang H, Chen T-H, Ting M-Y, Li X (2009) Efficient design-specific worst-case corner extraction for integrated circuits. In: Proceedings of design automation conference (DAC), pp 386–389

Chapter 5
High-Sigma Verification and Design

The Accuracy of Five Billion Monte Carlo Samples in Minutes

Abstract High-sigma IC designs are inherently difficult to create and verify. This chapter reviews various approaches for high-sigma analysis. It then describes High-Sigma Monte Carlo (HSMC), which is a high-sigma analysis approach that is fast, accurate, scalable, and verifiable. This chapter presents example results for representative high-sigma designs, revealing some of the key traits that make the HSMC technology effective. It describes how to extract full PDFs from -6 to $+6$ sigma, for application to statistical system-level analysis (e.g. for memory arrays). Finally, it presents industrial design examples.

5.1 Introduction

5.1.1 Background

High-sigma IC components can tolerate no more than a few defects in hundreds of millions or billions of instances. This is because these components tend to be replicated in large arrays, and therefore producing a single working product requires that a large number of the replicated components all work correctly. Common examples of high-sigma components are bitcells or sense amps in memory designs, and replicated digital standard cells. Some products, such as automotive or medical equipment parts, also have high-sigma requirements because circuit failure can have catastrophic consequences.

High-sigma parts are inherently difficult to design and verify because it is difficult to measure the effects of variation on high-sigma designs quickly and accurately. When there are only a few defects in a very large number of samples, Monte Carlo (MC) sampling requires a prohibitive amount of time to run in order to obtain accurate information in the extreme tail of the distribution where the

T. McConaghy et al., *Variation-Aware Design of Custom Integrated Circuits: A Hands-on Field Guide*, DOI: 10.1007/978-1-4614-2269-3_5,
© Springer Science+Business Media New York 2013

defects occur. Other methods, such as extrapolating results from a smaller number of MC samples or importance sampling, have other drawbacks such as long runtimes, poor accuracy, or are only effective for trivial examples and do not scale to the needs of production designs.

The result is that the actual sigma of high-sigma designs is often unknown, and additional margin must be added to compensate for this uncertainty. This in turn sacrifices power, performance, and area. Still, some designs may fail to meet their high-sigma goals, resulting in poor yields and expensive re-spins.

5.1.2 The Challenge

What is required is a fast, accurate, scalable, and verifiable approach for measuring high-sigma designs. We define these attributes as follows:

- *Fast:* Runs fast enough to facilitate both iterative design and verification within production timelines.
- *Accurate:* Provides information in the extreme tails of the high-sigma distribution, from real Monte Carlo samples.
- *Scalable:* Applicable to production-scale high-sigma designs with hundreds or thousands of process variables. Since modern, accurate models of process variation can have 10 or more process variables per device, typical rare-failure event memory circuits have 50–200 process variables, and digital circuits have thousands of variables.
- *Verifiable:* Results can be understood, pragmatically verified, and trusted. If the approach fails, it will be apparent to the user, just as a SPICE failure is indicated by non-convergence to DC operating point.

The High-Sigma Monte Carlo (HSMC) approach has been designed to meet the above requirements. This chapter presents an overview of existing high-sigma techniques, and then presents HSMC, including an overview of the technical approach and sample results on real production designs.

5.2 Building Intuition on the Problem

To start to get a feel for the problem, we simulated 1 million Monte Carlo (MC) samples for a 6 transistor (6T) bitcell, measured the read current, and examined the distribution. The bitcell has reasonable device sizings. Some simple yet popular models of local process variation, such as one ΔV_{th} per device, are not accurate (Drennan and McAndrew 2003). Modern models of statistical process variation are more accurate, having 5, 10, or more local process variables per device. For our examples, we use a 45 nm industrial process, with 5 process variables per device. The bitcell has 30 process variables total.

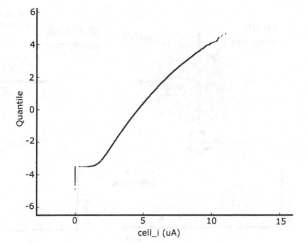

Fig. 5.1 NQ plot for bitcell read current, with 1M MC samples simulated

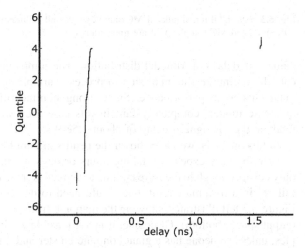

Fig. 5.2 NQ plot for sense amp delay, with 1M MC samples simulated

Figure 5.1 illustrates the distribution of bitcell read current (*cell_i*), in NQ plot form, where each point is an MC sample point. NQ plots make it easier to see the tails of a distribution, as Chap. 3 described in detail. In an NQ plot, the x-axis is the circuit output and the y-axis is the cumulative density function (CDF) scaled exponentially. In a circuit with linear response of output to process variables, the NQ curve will be linear—a straight line of points from the bottom left to the top right. Nonlinear responses give rise to nonlinear NQ curves.

In the bitcell NQ plot of Fig. 5.1, the bend in the middle of the curve indicates a quadratic response in that region. The sharp dropoff in the bottom left shows that for process points in a certain region, the whole circuit shuts off, for a current of 0. The curve's shape clearly indicates that any method assuming a linear response will be extremely inaccurate, and even a quadratic response will suffer.

Figure 5.2 shows the NQ plot for delay of a sense amp, having 125 process variables. The three vertical "stripes" of points indicate three distinct sets of

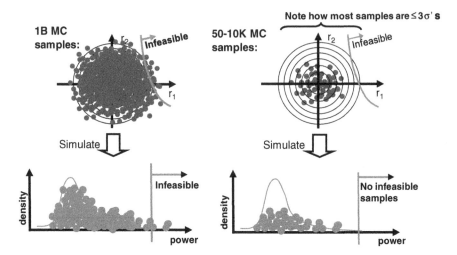

Fig. 5.3 *Left* 1 billion simulated MC samples *will* find failures on a high-sigma circuit. *Right* 10K simulated MC samples will not find failures

values for delay—a trimodal distribution. The jumps in between the stripes indicate discontinuities: a small step in process variable space sometimes leads to a giant change in performance. Such strong nonlinearities will make linear and quadratic models completely fail; in this case they would completely miss the mode at the far right at delay of about 1.5e-9 s.

In this analysis, we have shown the results of simulating 1M MC samples. But that can be very expensive, taking hours or days. Furthermore, 1M MC samples only covers enough to measure circuits to about 4 sigma. To find on average a single failure[1] in a 6-sigma circuit, one would need to do about 1 billion MC samples. Figure 5.3 left illustrates, showing the mapping from process variable space (top) to output space (bottom). Of course, it is not feasible to simulate 1 billion MC samples, unless someone has a giant compute cluster and a month of free time.

But of course, taking fewer MC samples means there will be no failures found, as Fig. 5.3 right illustrates. In this scenario, we have essentially no information about the high-sigma tails.

5.3 Review of High-Sigma Approaches

We aim to design and verify memory circuits and other circuits with rare failure events. Engineers and researchers have proposed a number of approaches for rare-event verification. To be applicable to production designs, an approach must be

---

[1] Where a failure is either failing a spec, or failing to simulate which also implies failing spec.

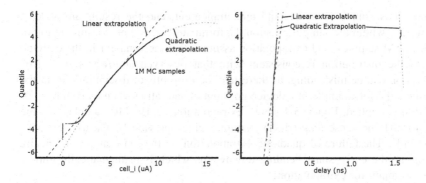

Fig. 5.4 Extrapolation from 1M simulated MC samples. *Left* bitcell read current. *Right* Sense amp delay

fast, accurate, scalable, and verifiable. This section summarizes some of the popular approaches and highlights challenges with these methods.

5.3.1 Giant Monte Carlo

As described in the previous section, MC would require hundreds of millions or billions of samples in order to produce a handful of failures for a high-sigma design; and simulating fewer samples means that no failures would be found. Positive attributes of MC include its quantifiable accuracy, results that are trustworthy, and scalability that is independent of dimensionality. This latter attribute is a truly remarkable property: MC accuracy is proportional to $1/\sqrt{N}$, where N is the number of MC samples; it is not related dimensionality or size of the circuit at all!

5.3.2 Medium MC

As described in the previous section, drawing 10K samples and simulating may be fast enough, but returns no information about the high-sigma tails.

5.3.3 MC with Extrapolation

This approach runs a large, but feasible number of MC simulations (e.g. 100K or 1M), then extrapolates the results to the region of interest. Extrapolation is typically done using curve fitting or density estimation. The benefits of this approach are that it is simple to understand and implement, and the results are at least trustworthy within the sampling region. Unfortunately, it is time-consuming to run 100K or 1M samples, and extrapolation assumes that the behavior in the extreme tails of the distribution is consistent with that observed at lower sigma. This assumption can be misleading, as there may be drop-offs or discontinuities in the extreme tails; for example, if a device goes out of saturation when a given level of variation is applied. Figure 5.4 shows extrapolation on 1M MC samples, for the bitcell (left) and sense amp (right) examples given previously. Clearly, extrapolation fails. The failure of quadratic extrapolation on the sense amp is tragically humorous: the curve starts bending downwards which is mathematically impossible. So much for extrapolation!

5.3.4 Manual Model

In this approach, one manually constructs analytical models relating process variation to performance and yield. However, this is highly time-consuming to construct, is only valid for the specific circuit and process, and may still be inaccurate. A change to the circuit or process renders the model obsolete.

The approaches described so far are traditional industrial approaches. The approaches that follow are more recent.

5.3.5 Quasi Monte Carlo

Also called "Low Discrepancy Sampling" (Niederreiter 1992), this variance-reduction technique draws samples from the process distribution with better spread, which in turn reduces the number of samples to get the same accuracy as MC. However, this does not solve the core problem of handling rare failures: for a 1-in-a-billion chance of failure, even a perfect QMC approach will need on average 1 billion MC simulations to get 1 failure.

5.3.6 Direct Model-Based

This class of approach uses models to evaluate a sample's feasibility far faster than simulation. The approach (Wang et al. 2009) adaptively builds a piecewise-linear model; it starts with a linear regression model, and at each iteration it chooses a higher-probability process point with known modeling error, simulates, and adds another "fold" to the model. The approach (Gu and Roychowdhury 2008) is similar, but uses a classification model rather than a regression model. With a model in place, one may do Monte Carlo sampling directly on it, importance sampling (Wang et al. 2009), or for some models analytically integrate for a yield calculation (Gu and Roychowdhury 2008).

The general problem of model-based approaches is that the model must be trustworthy, and there is no pragmatic method for verifying the accuracy of a high-sigma model; if the model is inaccurate, then the results will be inaccurate. Even if the model error is just 1 %, that 1 % error can translate directly to improperly labeling a feasible point as infeasible, or vice versa. Furthermore, these approaches have only been demonstrated on problems of just 6–12 variables; producing a reliable model for 60–200 variables is far more difficult.

5.3.7 Linear Worst-Case Distance

This approach (Schenkel et al. 2001) is a specific instance of the "Direct Model-based" approach, where each output has a linear model mapping process variables to performance (or feasibility). The n-dimensional model is constructed from $n + 1$ simulations: one simulation at nominal, and one perturbation for each of n process variables. Fig. 5.5 illustrates.

The biggest (and obvious) limitation is that these models are linear; whereas real-world high-sigma problems including bitcells are often highly nonlinear. As an example, consider the NQ plot in Fig. 5.1; the mapping would be linear only if the samples followed a straight line; but we see that the samples follow a quadratic curve, and in fact drop off on the bottom left when the transistor switches off (a very strong nonlinearity). Or in Fig. 5.2, a linear mapping will not capture the sharp discontinuities between the vertical "stripes". A linearity assumption can lead to estimates of yield that are dangerously optimistic.

Nonlinear WCD methods exist too; they are typically quadratic. Unfortunately, these approaches only assume a single region of failure, and scale much worse with the number of process variables.

5.3.8 Rejection Model-Based (Statistical Blockade)

This approach (Singhee and Rutenbar 2009) draws MC samples, but uses a classifier to "block" MC samples that are not in the 97th percentile tails. It simulates the remaining samples, and uses the ones beyond the 99th percentile to estimate yield or construct a tail distribution. The extra 2 percent is a safety margin to account for classifier error, which avoids the need for perfectly accurate models, via the safety margin. It avoids designers' possible distrust of sampling from alternate distributions by drawing directly from the process distribution. Figure 5.6 illustrates.

A problem is that the classifier model could easily have >2 % error (the approach has no way to guarantee this), which means it could inadvertently block

Fig. 5.5 Linear worst-case
distance (WCD) approach

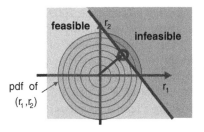

Fig. 5.6 Rejection model-
based (statistical blockade)
approach

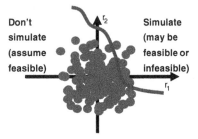

samples that are in the tail, and there is no effective method to detect this failure
condition. Furthermore, this method was demonstrated using problems with just
6–12 variables; not the 60–200 needed for industrial practice, which is far more
difficult to do.

5.3.9 Control Variate Model-Based

This variance-reduction technique uses the assistance of a "lightweight model" to
reduce the variance of the yield estimate. It combines the model predictions with
simulated-value predictions. Because of this, CV models have no minimum-
accuracy needs, unlike the above model-based approaches. Rather, CV approaches
merely get faster with models that are more accurate.

However, like QMC, control variates do not solve the core problem of handling
rare failures: if failure drops off only one-in-a-billion times, then CV will not be
able to help.

5.3.10 Markov Chain Monte Carlo

The MCMC approach recognizes that we do not need to draw samples directly
from the distribution; instead, we can create samples that are infeasible more often,
so that decent information is available at the tails. The MCMC approach derives

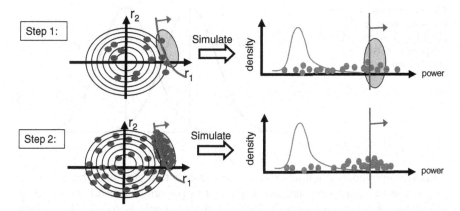

Fig. 5.7 Importance sampling. *Top* Step one, to find the region of failure (oval). *Bottom:*Step two, to sample in the region of failure

from the famous Metropolis–Hastings algorithm (Metropolis et al. 1953). In MCMC (Kanoria 2010), the sampling distribution adaptively tilts towards the rare infeasible events, and then stochastically uses or rejects each subsequent sample in the "chain", based on a threshold.

Unfortunately, a stable "well-mixed" chain of MCMC samples is difficult to achieve reliably in practice, especially for non-experts in MCMC (i.e. tool users). Even more troublesome is the arbitrariness of the sampling PDF: in real-world problems with dozens or hundreds of random process variables, it is difficult for users to gain insight into the nature of the sampling distribution, and therefore harder to trust the results or to know when MCMC may have failed.

5.3.11 Importance Sampling

In importance sampling (IS) (Hesterberg 1988), the general idea is to change the sampling distribution so that more samples are in the region of failure. It is typically composed of two steps, as shown in Fig. 5.7. The first step finds the new sampling region, which may be via uniform sampling, a linear/WCD approach, or a more general optimization approach. The step typically finds a "center" which is simply a new set of mean values for the sampling distribution. The second step continually draws and simulates samples from the new distribution; it calculates yield by assigning a weight to each sample based on the point's probability density on the original and new sampling distributions.

IS was first discussed for application to circuits in (Hocevar et al. 1983). Archetypical IS approaches for circuit analysis are (Kanj et al. 2006; Qazi et al. 2010), in which "centers" are computed and subsequently used in importance sampling, where the centers are the means of Gaussian distributions. To be more

Fig. 5.8 The mean of the
true distribution (p(x)) is
shifted by u_s so that more
samples are outside of
specification

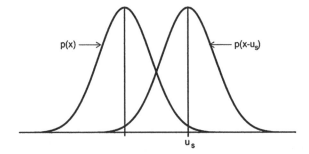

specific, importance sampling creates a new distribution g(x) such that a *greater
proportion* of samples are failures, compared to the original distribution p(x). It
then draws samples from g(x) and checks if the sample is *feasible* [i.e. meets
specifications, represented by $\theta(x)$], and then estimates yield by mathematically
unbiasing the samples.

To ensure a good proportion of samples that are *infeasible* (i.e. outside of
specifications), the centers-based IS shifts the mean of p(x) onto the failure
boundary, to get $p(x-u_s)$, where x is a "center". Figure 5.8 illustrates.

Various approaches apply different heuristics to find the "center" or mean
shift (u_s). The overall goal is to find the most-probable-point (MPP) that still
causes a failure. (As we shall see, failure-causing points that are not as likely hurt
the quality of the final estimation.) In the first-order reliability method (FORM)
(Hohenbichler and Rackwitz 1982), each parameter is perturbed one-at-a-time, in
order to construct a linear response surface, from which the MPP is calculated via
simple geometry. In the mixture-IS method (Kanj et al. 2006), ≈ 30 centers are
computed by drawing samples from a uniform distribution in $[-6, +6]$ standard
deviations for each process parameter, and keeping the first 30 infeasible samples.
Qazi et al. (2010) chooses centers via a spherical sampling technique.

Yield is calculated as follows. Each sample i has a *weight* ($w_i(x)$), which is a
function of the true distribution p(x) and the sampling distribution g(x):

$$w_i(x) = g(x)/p(x)$$

Normalized weights are:

$$v_i(x) = w_i(x) / \left(\sum_j w_j(x)\right)$$

Then, using the importance sampling *ratio estimate* (Hesterberg 1988), the
yield (Y) is:

$$Y = 1/N_s^* \sum_i (v_i(x)^* \theta_i(x))$$

where N_s is the number of samples, and $\theta_i(x)$ is 1 if x is feasible, and 0 otherwise.
$\theta_i(x)$ is determined via circuit simulation.

The IS approaches described above were demonstrated on circuit problems of 6–12 random process variables. But recall that for accurate industrial models of process variation such as (Drennan and McAndrew 2003), there are 5, 10, or more process variables per device. This means that even for a 6T bitcell, there are ≥ 30 process variables; and our sense amp problem has 125 process variables. We have seen industrial high-sigma analysis problems with 1,000 and even 10,000 process variables.

While IS has strong intuitive appeal, it turns out to have very poor scalability in the number of process variables, causing inaccuracy. Here's why: step one of IS needs to find the most probable points that cause infeasibility; if it is off even by a bit, then the average weight of the infeasible samples will be too low, giving estimates of yield that are far too optimistic. For example, in running (Kanj et al. 2006) on a 185-variable flip-flop problem, we found that the weights of most infeasible samples were <1e-200, compared to feasible sample weights of 1e-2 to 1e-0. This resulted in an estimated probability of failure of $\approx$ 1e-200, which is obviously wrong compared to the "golden" probability of failure of 4.4e-4 (found by a large MC sample). Sometimes a few samples "get lucky" and have larger weights than the rest, but that means that yield is being estimated from just a few sample points. One has no guidance at all on whether the approach has succeeded (found the global optimum) or failed (got stuck in a local optimum).

Reliably finding the most probable points amounts to a global optimization problem, which has exponential complexity in the number of process variables—it can handle 6 or 12 variables (search space of $\approx 10^6$ or 10^{12}), but not e.g. 30 or 125 as in the industrial bitcell and sense amp problems given before (space of 10^{30} or 10^{125}), let alone problems with 1,000 variables (space of 10^{1000}).

IS has another issue: because the sampling distribution is different than the true distribution, IS is harder for designers to trust and adopt.

5.3.12 Worst-Case Distance + Importance Sampling

This approach is a variant of importance sampling, where the "centers" are chosen in a style similar to the "Linear Worst-Case Distances" (WCD) approach, or local quadratic optimization variants.

Its issues are in accuracy and trustworthiness. Specifically, the linear version assumes that the slope at the nominal process point will lead to the most probable region of failure, which can easily be wrong, and leads to overoptimistic estimates of yield. The quadratic approach makes a local quadratic assumption rather than linear assumption, but is still susceptible to getting stuck on local optima and therefore getting overoptimistic yield estimates. Like IS, one has no guidance at all on whether the approach has succeeded or failed.

None of the approaches described so far meet all our goals: speed, accuracy, scalability, and verifiability.

5.4 High-Sigma Monte Carlo Method

5.4.1 An Idea to Break the Complexity Barrier

While importance sampling (IS) sounds promising, its reliability is hindered by the need to solve a global optimization problem of order 10^{30}–10^{125} (or 10^{1000}).

Perhaps we can *reframe* the problem and associated complexity, by operating on a *finite set of MC samples*. If we have 1B MC samples, then that is an upper complexity of 10^9. These are just 10^9 points to search across. While "just" 10^9 is much better than the 10^{125} complexity of IS, it is still too expensive to simulate 1B MC samples. But what if we were sneaky about which MC samples we actually simulated? Let us use an approach that prioritizes simulations towards the most-likely-to-fail cases. It never does an outright rejection of samples in case they cause failures; it merely de-prioritizes them. It can learn how to prioritize using modern machine learning, adapting based on feedback from SPICE. By never fully rejecting a sample, it is not susceptible to inaccurate models; model inaccuracy simply adds some noise to convergence, as we shall see later.

These are the central ideas behind the High-Sigma Monte Carlo (HSMC) approach.

5.4.2 HSMC Overview

High-Sigma Monte Carlo (HSMC) is a fast, accurate, scalable, and verifiable approach for verifying high-sigma designs.

HSMC works by generating a large number of MC samples, ordering the samples, then running the worst-case samples until all failures are found or until the extreme tails of the distribution are well established. This both generates a SPICE-accurate view of the extreme tail and enables an accurate prediction of the sigma value for the design.

Figure 5.9 shows the high-level algorithm of HSMC, which is summarized as follows. The engine inputs N_{gen}, the number of samples to generate. The algorithm draws N_{gen} samples from the process distribution. From these samples, it selects a subset of N_{init} samples and SPICE-simulates them. Assuming just one output, the algorithm constructs a model, the N_{init} points as training inputs, and the corresponding N_{init} performance values as training outputs. The candidate MC samples are from the N_{gen} MC samples, the ones not simulated yet. The algorithm simulates each point on the model to get predicted output values, then orders in ascending (or descending) order of output value. The algorithm then starts to simulate the samples in that order. Periodically, the algorithm will update the model with training data, and re-order the remaining candidate samples. The algorithm either stops manually when the user hits the "stop" button; or automatically when a stop criterion is hit, such as having detected all failures found, or ran 5,000 simulations.

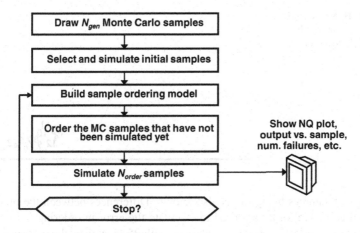

Fig. 5.9 High-level HSMC algorithm

5.4.3 Detailed Description

We now give a more detailed explanation of each of the steps.

Step: Draw samples: HSMC generates a large number of MC samples, enough to estimate yield to the target sigma level. The number of samples to generate is determined using a look-up table based on the target sigma value for the design. For example, 6 sigma analysis would use 5 billion generated samples.

Step: Select and simulate initial samples: From the MC samples generated, a subset is selected for use in building an ordering model. It selects the N_{init} (e.g. 1,000) samples that are farthest from nominal. This covers the extreme values in process variable space, while uniformly searching in the space of direction vectors from nominal. This subsampling method implicitly searches for directions of failures, with uniform bias to different directions. The subsample is then simulated using SPICE.

Step: Build sample ordering model: A regression model is then created using the patent-pending technology FFX (McConaghy 2011), which leverages advances in machine learning to handle arbitrary nonlinearities and high dimensionality. A separate ordering model is produced for each circuit specification.

Step: Order the samples: Using the ordering model generated in the previous step, the full set of MC samples generated in the first step is ordered, from the worst case to the best case. This is done for each specification. At this point, the predicted order of the samples, from worst to best, is known for each specification. Figure 5.10 illustrates.

Step: Simulate the predicted tails: The samples are then simulated using SPICE in predicted order, starting from the worst case. For each specification, the worst sample is run, then the second worst, and so on, until all failure cases are detected. The SPICE simulator provides the real value for each sample, and so the

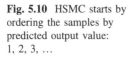

Fig. 5.10 HSMC starts by ordering the samples by predicted output value: 1, 2, 3, ...

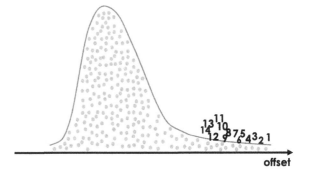

real order of samples being run is also known. Differences between the predicted order and the actual order are used to calculate the error in ordering.

Using this error, it is possible to calculate with 95 % statistical confidence when all failures to meet specification are found. The simulations stop when all failures are found. If the error in the predicted order is too great, the algorithm rebuilds the model with the new samples added, re-orders the remaining samples, and runs additional simulations. This iterative step can help to correct problems with the ordering model. Figure 5.11 illustrates this step.

Step: Show NQ plot, etc: Assuming that all failures are found, HSMC predicts yield (sigma) for a given spec value, or spec value for a given yield. It does this by assuming that all samples that were not simulated meet specification. Specifically, $yield = (N_{gen} - N_{fail})/(N_{gen})$, where N_{gen} is number of samples generated, and N_{fail} is the number of process points that failed to meet specifications. By sweeping different spec values, one can compute the spec value for a given target yield.

NQ plots are generated in a similar fashion; these plots show the tradeoff between yield (in units of sigma) and spec value.

HSMC also returns the 95 % confidence interval (lower and upper bounds) for yield value, using Wilson's score (Wilson 1927) for a binomial distribution. Wilson's score makes no assumptions about the shape of the distribution, it only uses the count of number failed and number generated.

5.4.4 Output of HSMC Method

HSMC produces, typically in hundreds or a few thousand simulations:

- An NQ plot, giving an accurate view of the extreme tail of the output distribution. Since this gives a tradeoff between yield and spec, one may get accurate yield estimate for a given spec; or spec estimates for a given target sigma (yield).
- A set of high-sigma corners, which can be subsequently used for rapid design iterations.
- A convergence curve, of output value vs. sample, which the designer can use to verify the convergence of the algorithm.

Fig. 5.11 HSMC simulates
samples in worst-case first
order 1, 2, 3, ..., until all
failures are found

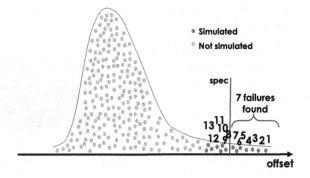

5.4.5 Example HSMC Behavior, on Bitcell Read Current

As HSMC runs, it is predicting which process points produce the most extreme
output values. If it does a perfect job of predicting, then on its very first simulated
sample it will get the very worst (maximum or minimum) value; on its second
simulated sample, it will get the second-worst value; and so on. One can plot the
output vs. sample value, and with perfect behavior the curve would be monotonically decreasing (if maximum-first) or monotonically increasing (if minimum-
first).

But of course since the model predictions are not perfect, the predicted order
may not be perfect—and that's ok! HSMC may make some erroneous guesses, but
with more simulations some of the guesses will be the worst-cases or near-worst
cases, and those are the most useful. This is the heart of the HSMC's robustness:
the model can have high error, but as long as the trend is from extreme value
inwards, then HSMC will find the tails.

Figure 5.12 compares typical HSMC output value vs. sample convergence to
"Ideal" and "Monte Carlo" behavior on a typical problem: bitcell read current
(*cell_i*). The detailed problem setup is given in Sect. 5.5, but here we focus on the
behavior. The aim is to find the samples with maximum values of *cell_i* first. In
this example, we have ideal data because we simulated all 1.5M of the generated
samples. The "Ideal" curve is computed by sorting the 1.5M *cell_i* values; the first
20K samples are plotted. We see that it monotonically decreases, as expected of an
ideal curve. The "Monte Carlo" curve is the first 20K simulations of MC sampling. As expected, it has no trend because its samples were chosen randomly; its
output values distribute across the whole range. So of course, MC is very slow at
finding the worst-case values; we can only expect it to find all worst-case values
once it has performed all 1.5M simulations.

The HSMC curve in Fig. 5.12 has a general downward trend starting at the
maximum value, with some noise in its curve. The trend shows that HSMC has
captured the general relation from process variables to output value. The noise
indicates that the HSMC model has some error, which is expected. The lower the
modeling error, the lower the noise, and the faster that HSMC finds failures.
At about 2,000 samples, the lower-range values for HSMC jump upwards. This is

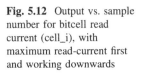

Fig. 5.12 Output vs. sample number for bitcell read current (cell_i), with maximum read-current first and working downwards

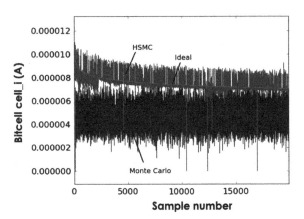

because HSMC has rebuilt its model using more data, and it has made the ordering more accurate. In a few sample points, such as at about 10,500 samples, HSMC predicted that some points would have extreme-maximum values, but when simulated they had extreme minimum values—that is acceptable because HSMC's success is not dependent on getting every sample predicted within an error tolerance. HSMC's success is based on quickly finding the worst case samples in a low number of simulations.

The HSMC curve of Fig. 5.12 provides transparency into the behavior of HSMC, to understand how well HSMC is performing in finding failures. This is a major part of the "verifiability" of HSMC: the user can tell if HSMC is having difficulty. The width of the noise shows how much margin should be given prior to concluding that all failures have been found for a given specification value. The clear trend shows that HSMC is working correctly and is capturing the tail of the distribution.

5.4.6 HSMC Usage Flows

HSMC is efficient both for verifying high-sigma designs and for finding high-sigma design corners. As such, it is useful not only for verification, but also within the design loop. This provides much more opportunity to tune designs, reducing the need to over-margin, and producing more predictable results when verifying.

5.4.6.1 Sizing Circuits with the Help of HSMC

We first review the "generalized" corner concept discussed in earlier chapters, then describe its application to high-sigma design.

The idea of corners has been around for a long time, and used broadly in many classes of circuit design including memory. Typically, one thinks of corners as

Fig. 5.13 A six-sigma sizing
flow

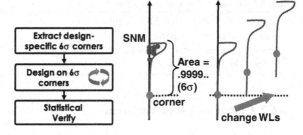

PVT corners: a model set value such as FF or SS, and environmental conditions like voltage and temperature. The idea behind corners is sound: find some representative points that bound the distribution of performance, and design against those. This enables fast design iterations without having to do a full statistical analysis at each candidate design. FF/SS corners bracket digital device performance fairly well, which propagates to digital circuit performances of speed and power fairly well (at least traditionally). However, FF/SS corners do a poor job of bracketing the distributions of memory performances.

Let's take the idea of corners—bracketing performance for rapid design iterations—and make it more general than PVT corners in order to *accurately* bracket memory performance distributions. Figure 5.13 illustrates the flow. (This is the sigma-driven flow introduced in Chap. 4, but now the application is to high-sigma design.)

The first step is to *extract* 6-sigma corners. This is done by simply running HSMC, and selecting the process point with an output performance value closest to 6 sigma.

In the next step, bitcell/sense amp designs with different candidate sizings are tried, using whatever methodology the designer prefers. For each candidate design, the user only needs to simulate on the corner(s) extracted in the first step. The output performances are "at 6-sigma yield". The distribution of the output (SNM[2] in the figure) is *implicitly* improved. This allows exploration of tradeoffs among different performances, at 6-sigma yield, but only having to do a handful of simulations (one per corner).

In the final step, the designer verifies the yield by doing another run of HSMC. The flow concludes if there was not significant interaction between process variables and outputs. Sometimes there is significant interaction, in which case a re-loop is done: grabbing a new corner, designing against it, and verifying. Typically, only one re-loop at most is needed because the design changes in the re-loop are smaller.

---

[2] SNM = static noise margin.

Table 5.1 Nested MC for global + local

For each of 100–1,000 global MC samples
    Draw a global process point
    Run MC: For each of 1M-1B local MC samples, draw a local process point and simulate
        netlist at {global, local}.

5.4.6.2 Global vs. Local Variation

Here, we discuss how global (die-to-die, wafer-to-wafer) reconciles with local (within-die) statistical process variation in an HSMC context. We consider five different approaches.

Nested MC: This is given in Table 5.1. An outer loop simulates different wafers or dies being manufactured, and an inner loop simulates the variation within the die. The approach is simple, and handles all nonlinear responses and interactions among global and local. But of course, it is far too slow, needing billions of simulations.

No global + Local HSMC: The idea here is to simply ignore global variation, by setting its variables to nominal; then to run HSMC with local variation. It is simple, fast, convenient, and actually has many good use cases. But of course it ignores global variation.

Global FF/SS Corner + Local HSMC: The idea here is to set global variation to a digital modelset value such as FF or SS. This is also simple, fast, and convenient, but is not a fully accurate reflection of the effect of global variation.

Global 3-Sigma Performance Corner + Local HSMC: First, extract a 3-sigma corner on global variation, e.g. using a "Monte Carlo" tool, which gives a process point with output value at the 3-sigma percentile in performance space. Then, run HSMC where the global variation is set to the values of the 3-sigma corner.

This is simple, fast, and convenient, and a much better reflection of the effect of global variation. It handles nonlinear responses to global variation and to local variation, and interactions between the global process *point* and local variations. Its relatively minor drawback is that it assumes that local variation does not affect the choice of global process corner. This is a safe assumption for getting a {global + local} process for rapid design iterations, but may not be as safe for a final verification step.

Nested HSMC: This approach, given in Table 5.2, is just like nested MC, except the inner MC loop is replaced by HSMC. The approach is simple, and handles all nonlinear responses and interactions among global and local. Its drawback is that it takes about 100x more simulations than a single HSMC run. However, given that typical HSMC runs are about 1,000 simulations, then 100K simulations are often justifiable for a final detailed verification step.

The previous section described how HSMC is used in the context of a rapid-iteration design flow, via corner extraction. This flow reconciles with global + local variation, as Fig. 5.14 illustrates. The first step, of corner extraction,

Table 5.2 Nested HSMC for global + local

For each of 100–1,000 global MC samples
   Draw a global process point
   Run HSMC across local, using global process point just drawn.

Fig. 5.14 A 6-sigma sizing
flow, that reconciles global
variation

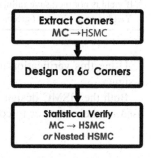

uses the "Global 3-Sigma Corner" approach. The final verification step uses
"Global 3-Sigma Corner" too, or "Nested HSMC", depending on the user's time
constraints vs. design aggressiveness.

5.4.6.3 HSMC for High-Sigma Verification

For verifying designs, it may be desirable to measure accuracy more precisely,
while extending runtime in order to do so. Table 5.3 summarizes a methodology to
do so.

Step 1 of Table 5.3 is a pilot run to verify the ordering model accuracy by using
a smaller number of samples (e.g. 10K). Configure HSMC to run all 10K samples,
such that the entire distribution is known. Next, compare HSMC's predicted order
with the actual order throughout the distribution using the plot included in the
HSMC app. As long as there is a very strong trend in the predicted order in the
worst-case tails of the distribution, and there is not a significant number of failures
found outside of the predicted worst-case tails of the distribution, HSMC's
ordering model can be verified to be strong for the design in question, at least at
lower sigma values.

Step 2 is for yield estimation, using HSMC if step 1 showed it to be accurate,
and the prior status quo method if not (e.g. MC with extrapolation). The limit on
samples simulated should be set to a higher value (e.g. 20K), taking into account
project schedule. HSMC will stop sooner if it detects that all failures are found.

In step 3, if the design fails verification, then the corners found can be saved and
designed against, making the result of the next verification attempt more
predictable.

Table 5.3 High-sigma verification flow

1. Test HSMC accuracy on the circuit, by running 10K generated and 10K simulated samples. If HSMC shows a strong trend from worst case to best case, the ordering model is strong for this design.
2. If HSMC effectively models the design, run HSMC for verification, with xM/G generated and xK simulated samples. If not, run previous method (e.g. extrapolating 1M MC samples).
3. If yield is not met, save failures as corners and fix design using DesignSense, etc. Go to step 2.
4. Yield is met, so stop (success).

Using this verification methodology, HSMC can provide quality results up to 6-sigma.

Both the HSMC design flow and the HSMC verification flow methods can be used on their own, or to complement other high-sigma techniques.

5.4.7 Other HSMC Attributes

PVT Corners: HSMC handles PVT corners in a similar fashion to handling global variation. Just as it can take in a specification of 3-sigma global corners, it can take in a specification of a PVT corner. Just as one can use a nested HSMC configuration to loop on global MC samples, one can also loop on PVT corners.

RC Corners: RC parasitics in post-layout netlists can be treated similar to PVT corners.

Parallel processing: HSMC is designed to work efficiently whether there is 1, 10, 100, or 1,000 cores or machines, by leveraging parallel processing such as LSF and SGE. It uses parallel processing for simulation. Also, because sorting billions of samples can take several hours on one core, HSMC parallelizes that as well. Typically, speedup due to parallelization is nearly perfect, e.g. 10 cores runs about 10x faster and 100 cores about 100x faster.

Failed simulations: If the simulator fails to converge, or a measurement returns a NaN, HSMC takes note rather than simply ignoring it. These points really matter, because they may indicate a real circuit failure, and if not a real failure they need further investigation by the designer. HSMC handles failed simulations as follows:

- Before constructing the ordering model that maps process point values to output values, HSMC by default converts each failed simulation or NaN to a real number that represents "poor" behavior. Therefore the ordering will prioritize NaN numbers alongside real-valued "poor" output values.
- When yield is estimated, the failed simulations/NaNs are all treated as process points that fail specifications.
- In output vs. sample plots, failed simulations/NaNs are shown as points along the bottom or the top of the curve. Figure 5.15 illustrates.
- In plotted NQ curves, the failed simulations/NaNs are ignored, because there is no meaningful way to incorporate them.

Fig. 5.15 An output-vs.-
sample curve (max-first),
where the points along the
bottom indicated failed
simulations

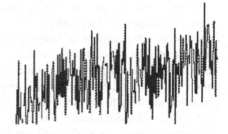

5.5   HSMC: Illustrative Results

5.5.1   Introduction

This section examines HSMC's behavior on a suite of designs. The purpose of this section is to show how HSMC works in practice on actual designs, and purposely includes both cases where HSMC works very effectively and cases where HSMC is less effective.

We show HSMC behavior on five different high-sigma problems: three circuits, one with a single output and two that have two outputs each. We test on a bitcell and a sense amp, which are representative memory circuits, and a flip-flop, which is a representative digital standard cell. The circuits have reasonable device sizings. The device models used are from a modern industrial 45 nm process, having approximately 5–10 local process variables per device. The bitcell has 30 variables, the sense amp has 125 variables, and the flip-flop has 180 variables.

5.5.2   Experimental Setup

The experimental methodology is as follows. For each problem, we drew $N \approx$ 1M Monte Carlo samples and simulated them. These form our "golden" results. We set the output specification such that 100 of the N samples fail spec. Then we ran HSMC on the problem, with $N_{gen} = N$, using the same random seed so that it has exactly the same generated MC samples. HSMC ran for 20K simulations. We repeat the procedure with specs set such that 10 of the N_{gen} samples fail spec. $N = 1.5$M MC samples for the bitcell, 1M for the sense amp, and 1M for the flip-flop.

5.5.3   Bitcell Results

Section 5.4.5 first described the output vs. sample behavior of HSMC, on the bitcell read current (*cell_i*). For the reader's convenience, we repeat the convergence plot here, in Fig. 5.16. To summarize, we see that the HSMC convergence curve has a general downward trend; and that its noise is perfectly acceptable.

HSMC's effectiveness in finding failures depends on the target specification. A correct setup would typically include fewer than 100 failures to meet specification within the number of samples generated. If there are more failures, then either the design is not meeting its target sigma, or there were too many samples generated for the target sigma. Similarly, if there are no failures to meet specification, then either the design is over-margined or there were not enough samples generated to verify to the target sigma. Therefore, HSMC only needs to be able to find up to a hundred failures to meet specification, allowing a tolerance for significant ordering error while still working within acceptable simulation budgets. In the bitcell case, HSMC finds the first 100 failures within its first 5,000 predicted samples (see Fig. 5.17). Note that with 1.5 million samples containing 100 failures, simple MC sampling will typically not find a single failure within 5,000 samples, as the bottom curve in Fig. 5.17 illustrates.

The bitcell results demonstrate one of the key strengths of HSMC, which is its resilience to order prediction error. Since the context only requires finding up to 100 failures to meet specification, the ordering model does not need to be perfectly accurate in order to deliver MC- and SPICE-accurate results in the extreme tails of a high-sigma distribution within a reasonable number of simulations.

5.5.4 Sense Amp Results

Figures 5.18 and 5.19 show HSMC behavior on the sense amp's power output. Its behavior is similar to those seen in the bitcell, but it finds all 100 failures within its first 1,000 samples (see Fig. 5.19). The effect of the ordering model can be seen in Fig. 5.18, as the amount of noise shown in the HSMC curve is clearly lower relative to the sampling region. The sense amp power example illustrates how HSMC gains efficiency with a better ordering model.

Figures 5.20 and 5.21 show HSMC performance on the sense amp's delay output. This output has a bimodal distribution, with most sample values being about 0.1e-9 s, and failure cases having a value of about 1.5e-9. We set the spec in between; of the 1 million MC samples, there are 61 failing samples (rather than 100). Figure 5.20 shows that HSMC finds all failures within its first 9,000 samples. HSMC's behavior is to find failures with highest frequency in the earlier samples, with decreasing frequency. We can see visually on the output vs. sample plot in Fig. 5.20 that the ordering model is good because the frequency of failures is high at first, then drops off. We can also see that all failures are likely found because there are no new failures found over a large range of samples (i.e. from sample #9000 to #15000). Figure 5.21 further demonstrates this behavior; we see that HSMC finds all 61 failures within 9K samples, and that it finds most of the failures within the first 1,000 samples.

In the case of a bimodal output distribution, HSMC's behavior is still essentially the same as in the previous cases, though the noise propagates in a different manner. In this case, the upward trend in the output vs. sample number plot is

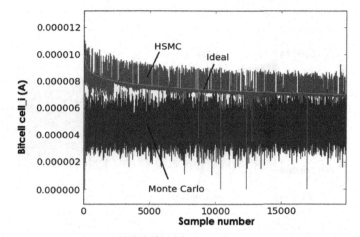

Fig. 5.16 Output vs. sample number for bitcell read current (cell_i), with maximum read-current first and working downwards

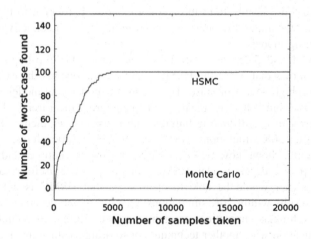

Fig. 5.17 Bitcell cell_i—number of failures found vs. sample number (100 failures exist)

replaced with a view of frequency of failures. Note that the frequency of failures drops off as more samples are run, and that it becomes clear that it is unlikely to find additional failures beyond 10K samples.

5.5.5 Flip-Flop Results

Figures 5.22 and 5.23 show HSMC's behavior on the flip-flop's V_h output. We see that HSMC performs near-ideally in output vs. sample convergence, and HSMC finds 100/100 failures in less than 500 samples.

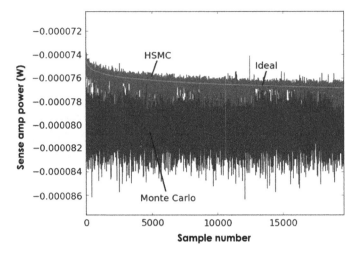

Fig. 5.18 Sense amp power—output vs. sample number (max-first)

Note again how visibly tight the amount of noise is relative to the sampling region. Again, the amount of noise is a good indicator of the effectiveness of the sample ordering model.

Figures 5.24 and 5.25 show HSMC's behavior on the flip-flop's I_d output. Figure 5.24 shows that while the HSMC curve biases towards the extreme maximum, it has a high degree of noise. This means the underlying model is capturing the global trend, but it has significant error in capturing local trends. Despite this significant error, it is still finding failures with reasonable efficiency. Figure 5.25 shows that after 20K simulations, HSMC has found 26/100 failures.

This example shows how HSMC is self-verifying at runtime, and how even with a very poor ordering model, HSMC can still produce useful results. In this case, the designer would be able to clearly see that HSMC is not producing dependable results within 20K simulations. Given this, the designer could opt to either run additional simulations to gain more resolution, to complement the HSMC verification with another technique, or to design with some added margin to account for the uncertainty. The designer can also use high-sigma corners discovered here to design against in a subsequent iteration. The key is that HSMC is not misleading due to the inherent quality that it is largely self-verifying.

5.5.6 HSMC vs. Extrapolated MC

5.5.6.1 Introduction

One common method for verifying high-sigma designs is to take a large number of MC samples, then extrapolate the output distribution to estimate the extreme tails. The main problems with this method are:

Fig. 5.19 Sense amp power—number of failures vs. sample number (100 failures exist)

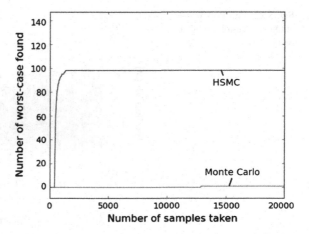

Fig. 5.20 Sense amp delay/1e9—output vs. sample number (max-first)

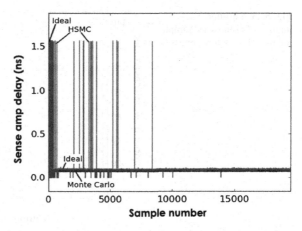

Fig. 5.21 Sense amp delay—number of failures vs. sample number (61 failures exist)

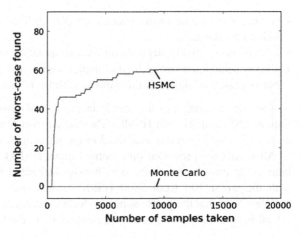

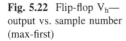

Fig. 5.22 Flip-flop V_h— output vs. sample number (max-first)

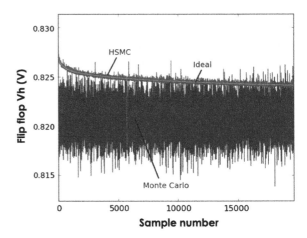

Fig. 5.23 Flip-flop V_h— number of failures vs. sample number (100 failures exist)

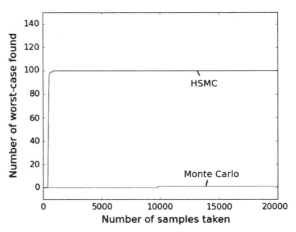

- It takes a long time to run enough samples to have a good chance at extrapolating into the tails.
- Extrapolation quality depends on the extrapolation technique chosen.
- Circuit behaviours can change at higher sigma, and so the assumption that the behavior at low sigma extrapolates gracefully is inherently risky.

This section compares the speed, accuracy, and verifiability of extrapolating 1 million MC samples with HSMC. These experiments use the same bitcell, sense amp, and flip-flop circuits examined in the previous section.

All results are presented on a normal quantile (NQ) plot to facilitate extrapolation. The plots compare the distributions estimated by 1 million MC samples with the worst 100 from 5500 HSMC simulations on 100 million generated samples. Note that the points appear to form lines due to their density, though they are all in fact individual points representing individual simulations.

Fig. 5.24 Flip-flop I_d—output vs. sample number (max-first)

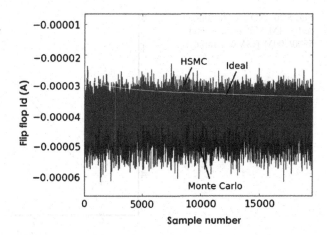

Fig. 5.25 Flip-flop I_d—number of failures vs. sample number (100 failures exist)

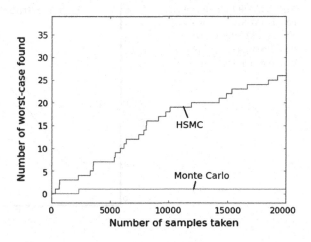

5.5.6.2 Extrapolation Cases

The results shown in Figs. 5.26, 5.27, 5.28, 5.29, 5.30, 5.31 and 5.32 demonstrate three main extrapolation cases.

MC results that extrapolate well: For the cases in Figs. 5.28, 5.29, and 5.31, MC extrapolates well and is an effective predictor of the extreme tails of the distribution.

MC results that extrapolate questionably: In the case shown in Fig. 5.26, MC provides an idea of what the extreme tail looks like, although the curve at the end of the MC data does not suggest a definitive extrapolation. In this case, any kind of prediction made will have some amount of error, which will depend on the extrapolation technique used. For example, Fig. 5.27 shows the same data, but with linear and quadratic extrapolation; both extrapolations capture the tail poorly.

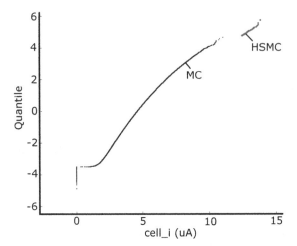

Fig. 5.26 NQ plot for bitcell
cell_i: 1M MC samples and
5500/100M HSMC samples

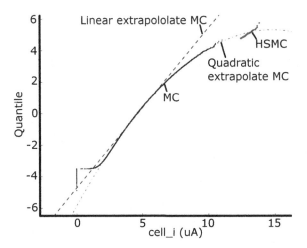

Fig. 5.27 NQ plot for bitcell
cell_i: 1M MC samples and
5500/100M HSMC samples,
with linear and quadratic
extrapolation curves

Similarly, for the case of Fig. 5.32, the MC data does not provide a clear indication of how to extrapolate. Note that the HSMC results in Fig. 5.32 are off: HSMC had missed some failures, and an NQ plot needs all failures to have correct x-axis values. This inaccuracy would be revealed to the user by the amount of noise shown in the output vs. sample number plot (see Fig. 5.24).

MC results that do not estimate the extreme tails: Some MC results simply do not serve as an effective estimator of the extreme tails. For example, consider the bimodality shown in Fig. 5.29. Imagine if, instead of 1 million samples, only 100,000 samples were run and the second mode was not revealed. Figure 5.30 illustrates this. It would not be possible to even know about the second mode, much less to extrapolate from it. The challenge here is that there is no way to know if this type of case will happen, and similarly no way to know how many MC samples need to be run in order to estimate it. HSMC captures these cases.

Fig. 5.28 NQ plot for sense amp power: 1M MC samples and 5500/100M HSMC samples

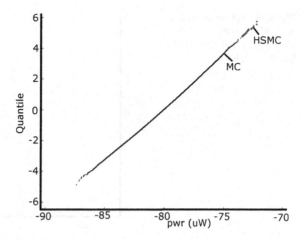

Fig. 5.29 NQ plot for sense amp delay: 1M MC samples and 5500/100M HSMC samples

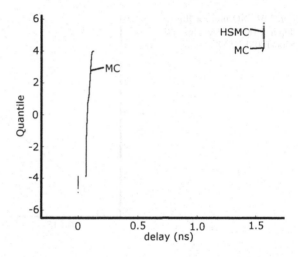

In summary, extrapolating MC results is best reserved for cases where HSMC produces very high noise in its output vs. sample plot, such as in the flip-flop I_d case. Otherwise, HSMC is faster because it requires fewer simulations, more accurate because it produces results in the extreme tails, and more verifiable because it self-verifies its model of the extreme tails at runtime.

5.6 Binary-Valued Outputs and Adaptive Initial Sampling

This section describes a particular issue that arises in some circuits—binary-valued outputs with rare failure modes; and a variation of HSMC that addresses it.

Fig. 5.30 NQ plot for sense amp delay: 100K (not 1M) MC samples and 5500/100M HSMC samples

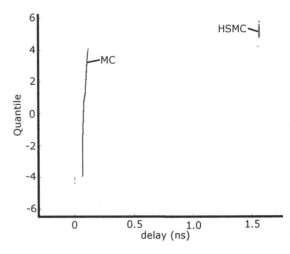

Fig. 5.31 NQ plot for flip-flop V_h: 1M MC samples and 5500/100M HSMC samples

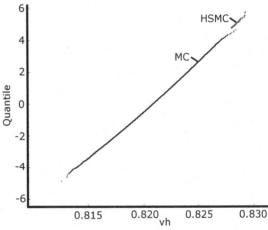

Fig. 5.32 NQ plot for flip-flop I_d: 1M MC samples and 5500/100M HSMC samples

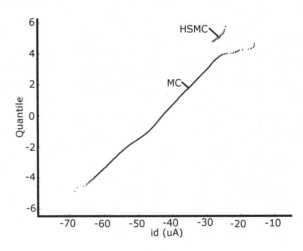

5.6.1 Default HSMC and Typical Successful Behavior

Let us label the HSMC variant, as described in the previous sections, as 'default HSMC'. In default HSMC, the initial samples are selected from the large set of generated MC samples.

Figure 5.33 shows the behavior of default HSMC, on a problem where HSMC behaves as expected and with success. The figure has emphasis on the initial samples, from the perspective of process-variable space. In step (a), the initial samples (dark dots) are chosen as the farthest-from nominal of generated MC samples (all dots). Those samples are simulated. In step (b), a model mapping initial input process-variable values to simulated output value is constructed. It is shown as the contour lines. The top-right shaded region is the true infeasible region, where any samples would not meet the output spec. It is up to HSMC to prioritize simulations towards points in this region. In step (c), the samples are ordered using the model, then simulated in that order. We see that this simulation order is correct, because it will focus on simulating the generated samples in the infeasible region.

5.6.2 Default HSMC on Binary-Valued Outputs

We now examine behavior of default HSMC on a specific circuit problem that causes difficulty. In this difficulty-causing problem, the circuit outputs just one of two values (binary-valued outputs), and the failing output is very rare (e.g. happens 1 in a billion times). Such problems can exist in practice; for example on a bitcell, rather than measuring read current, one measures "was there a read failure?"

Figure 5.34 shows default HSMC behavior on such a problem. In step (a), the initial samples (dark dots) are chosen as the farthest-from nominal generated MC samples, and simulated. But now, the simulated output value for every single initial sample is the same. Step (b) constructs a model of process variables to output values, but since all the output values are constant, the model is simply a constant too. Therefore, on Fig. 5.34b, there are no contour lines. The top-right shaded region is the true infeasible region, where any samples would not meet the output spec. On this circuit problem, this is the region that would return the *other* binary-valued output value. It is up to HSMC to prioritize simulations towards points in this region. In step (c), the samples are ordered using the model, then simulated in that order. Since the model is flat, the model-based ordering is not meaningful, and the ordered simulations will miss the samples in the infeasible region.

This issue happens in 'default HSMC' on a specific set of problem types:

- It can occur on binary-valued outputs where one output value is extremely rare, as described above.

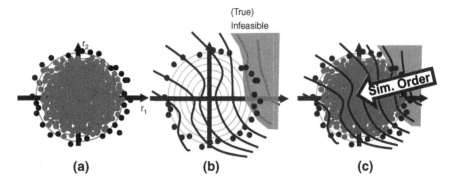

Fig. 5.33 Default HSMC, where initial samples are a subset of generated MC samples; HSMC behaves as expected. **a** The dark dots are initial samples in process variable space; all dots are generated MC samples. **b** The model was built from initial samples' simulation data; the contour lines are its predictions. **c** Simulations are ordered according to the model predictions/contours

- It can also happen in a slightly more general case, where there is a sharp dropoff in the output value (discontinuity) in the rare failure region, and contours of poor output values in the non-failing region do not lead to rare failure regions.

We discuss below how these specific problem types are addressed with either 'default HSMC', or with 'adaptive initial sampling' HSMC.

To be clear, 'default HSMC' *does* handle the following cases just fine.

- Default HSMC behaves fine with binary-valued outputs where both output values have more than negligible probability. In this case, default HSMC's initial sampling will find examples of both output values, and therefore construct a model that orders samples well.
- Default HSMC handles sharp dropoffs in output values, when contours of poor output values in the non-failing region *do* lead to rare failure regions. In this case, HSMC's ordered samples will find their way to failures, even if there were no failures in the initial samples. In practice, this is quite common.
- Default HSMC handles sharp dropoffs in output values, when contours of poor output values in the non-failing region do not lead to failure regions, but the failure regions are *less* rare. In this case, default HSMC's initial sampling will find failing samples, and therefore construct a model that orders samples well.

We now describe how 'default HSMC' can still address most difficulty-causing problem types, with a little more work on the part of the user. In practice, we have found that users assess whether or not HSMC has found failures in the initial sampling. If it *has* found failures, they let HSMC proceed with ordered sampling. But if it has *not* found failures, then they increase the number of generated samples or the number of initial samples, and run HSMC again. With a larger number of generated samples, the initial samples are farther away from nominal and often

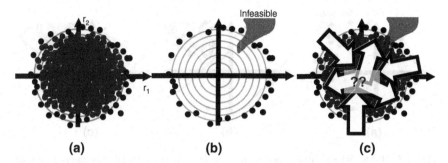

Fig. 5.34 Default HSMC, where initial samples are a subset of generated MC samples; but the outputs are binary-valued with rare failure cases, causing HSMC to miss the failing points. **a** The dark dots are initial samples; all dots are generated MC samples. Each initial sample has the same simulated output value. **b** There are no contour lines because the model has a flat response. **c** With no model contours, the ordered simulations miss the samples in the infeasible region

have a higher chance of finding failing samples. With a larger number of initial samples, the user is sampling more directions, increasing the chance of finding failing samples.

5.6.3 HSMC with Adaptive Initial Sampling on Binary-Valued Outputs

After some experience with the workarounds of more generated samples or initial samples, we developed a technique that *directly* handles the difficulty-causing problem types without the need for user iterations. The general idea is to find failures *adaptively* in initial sampling, rather than choosing and simulating a fixed set of generated MC samples. These revised initial samples do not even need to be from the generated MC samples.

Figure 5.35 illustrates behavior of HSMC, revised to have adaptive initial sampling, on binary-valued outputs. Step (a) is the initial sampling step. HSMC performs adaptive initial sampling until adequate failing samples are found. With non-failing and failing samples in hand, HSMC can build an ordering model. Step (b) of Fig. 5.35 shows the contours of the ordering model, which has correctly predicted different output values for the infeasible region. Finally, the generated MC samples are ordered according to the model predictions/contours; we can see that it correctly simulates the failing samples first.

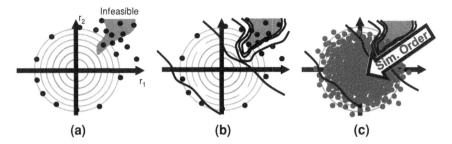

Fig. 5.35 HSMC, where initial sampling adapts to find failures. **a** Initial sampling is done until there are adequate failures, i.e. black dots in the shaded infeasibility region. **b** The model contours capture the feasibility boundary. **c** The generated MC samples are simulated in the correct order—with failing samples first

5.6.4 Details of Adaptive Initial Sampling Algorithm

This subsection describes exactly how adaptive initial sampling is performed. Note that this is more detail on adaptive initial sampling than many readers would care about; so it can readily be skipped.

The first-cut goal of the initial sampling algorithm is simply to find failing samples. In early experiments, we found that this was insufficient on its own. If the failing samples found were deep inside the infeasibility region, too far from nominal, they would affect the ordering model but not in a region that mattered because there were no generated MC samples that far out. The failing samples needed to be close enough to nominal (high enough probability) to affect the ordering model's prediction on generated MC samples. Cast as an optimization problem, this is: find the point(s) that minimize distance to nominal, subject to being infeasible.

The approach to solve the optimization problem has two phases. The first phase finds one or more failing samples, using "spherical sampling". The second phase applies local optimization to each failing sample, to get it closer to nominal while still failing.[3] Figure 5.36 illustrates.

Figure 5.36a–c illustrates the first phase, which performs spherical sampling adaptively. First, N samples are drawn as shown in Fig. 5.36a. Each sample has distance d from nominal, but the samples have uniformly spread direction vectors from nominal. Each of these samples is simulated. If the number of failures found so far is less than the target number of failures N_{thr}, then the distance d and the number of samples N are increased and the sampling is repeated. We see in

---

[3] Perceptive readers may see that a similar optimization problem exists in importance sampling (IS); and that the spherical sampling phase bears resemblance to the IS technique (Qazi et al. 2010). However, the problem for HSMC is easier than IS, because as Sect. 5.4.1 describes, HSMC only needs these points to influence its ordering of generated MC samples, rather than IS needing to settle on the choice of sampling region(s).

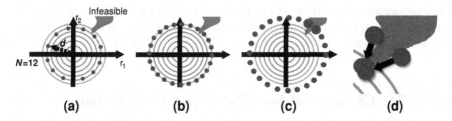

Fig. 5.36 Details of adaptive initial sampling. (**a–c**) The first phase did three iterations of spherical sampling, finally finding two failing points (**d**) The second phase minimized the distance to nominal, for each failing point

Fig. 5.36a that the samples did not hit the infeasible region, i.e. there were no failures. So, d and N were increased, and spherical sampling was repeated in Fig. 5.36b. After one more iteration, the sampling of Fig. 5.36c found enough failures to proceed to the second phase. N is typically a value of 100–1,000, and d is typically a value of 5–20 normalized standard deviations from nominal.

Figure 5.36d illustrates the second phase. Each of the two failing points was optimized to be as close as possible to nominal while still failing, using SPICE-in-the-loop. It has two steps:

1. **One-dimensional bisection search** between nominal process point and the failing sample, then
2. **n-dimensional local optimization.** At the each iteration, it draws a set of "child" candidate points from the "parent" best point so far. It replaces the parent with any child that is infeasible with lower magnitude. It generates children with these operators: Gaussian perturbation about parent; zero some values; halve some values; and model-building optimization. Model-building optimization builds a logistic regression classifier (Hastie et al. 2009) from nearby samples, and optimizes on the model in a restricted "trust region" (Celis et al. 1985) to find the best point.

This concludes our description of the adaptive initial sampling approach, which HSMC uses to handle binary-valued outputs with rare failure cases.

5.7 System-Level Analysis and Full PDF Extraction

5.7.1 Introduction

This subsection describes the challenge of statistical system-level analysis, including memories and large digital systems. It then reviews various approaches that have been proposed, with a focus on approaches for SRAM system analysis. It turns out that the approach of mixing PDFs of circuit blocks' performances is fast, accurate, and easy to apply. The singular challenge in that approach is to generate full performance PDFs from −6 sigma to +6 sigma, so we describe how HSMC is extended to extract full PDFs. This subsection concludes with an example analysis of an SRAM array.

5.7.2 The Challenge of Statistical System-Level Analysis

The discussion so far has considered high-sigma analysis where simulation time of the circuit is low enough to perform a few thousand simulations in a practical amount of time. However, consider the case of a whole SRAM memory slice having thousands of bitcells, where a single simulation takes several hours on modern Fast-SPICE simulators. Or, consider the case of very large digital circuits, which need hours to days of simulation time.

For such system-level circuits, it is possible to perform a handful of simulations on the system, but not enough for an accurate statistical analysis. Yet statistical analysis is extremely useful, because it will show the tradeoff among yield, timing, power, and area *at the system level*, making it easier to measure the SoC value proposition.

5.7.3 Brief Review of System-Level Approaches, Focusing on SRAM

Because of the usefulness of system-level statistical information, technologists have explored many ways to do practical analysis of SRAM systems and other systems. To give the reader a feel for the challenge, we review some approaches used in industry or proposed in the literature, for analyzing read current in SRAM systems.

- *Use worst-case values to set system-level performance.* This approach computes the worst-case read current of the bitcell (e.g. 6-sigma value) and worst-case offset voltage of the sense amp (e.g. 5-sigma value) via high-sigma analysis. These worst-case values are used to directly estimate system-level performance. However, there is an exceedingly rare chance that the very worst sense amp will be share the same bitline with the very worst bitcell. This means that the estimated system-level performance values will be exceedingly pessimistic.
- *MC sample whole slice, with simulation.* This approach draws 100–1,000 MC samples of the system, then simulates them. Clearly, this approach is too slow.
- *Nested MC on sense amp and bitcell process points, with simulation.* This approach has two loops. The outer loop draws MC samples of the sense amp in process variable space. The inner loop draws MC samples of the bitcell in process variable space. In one variant, each {bitcell, sense amp} is simulated as

a single netlist, connected in a bitline with the capacitance of thousands of bitcells. In another variant, the bitcell and sense amp are simulated separately. An outer MC sample passes if each bitcell has enough current to create a sufficiently large bitline voltage for the sense amp of that MC sample. This approach is slow, because it requires millions of simulations of the bitcell and sense amp.

- *Nested MC on sense amp and bitcell process points, with behavioral modeling.* This approach (Wang et al. 2009) has two steps. Its first step extracts a behavioral model for the sense amp, and for the bitcell. The next step is just like the previous approach, except now simulations are on the behavioral models. The issue is that behavioral model extraction is complex, and not very mature.
- *Loop-flattening MC on sense amp and bitcell process points, with simulation.* This approach (Qazi et al. 2010) employs statistical "law of large numbers" theory to reframe a nested Monte Carlo problem into a single flattened MC problem. It then solves the problem using a high-sigma analysis technique. Unfortunately, in our experience we have seen that this problem reframing is not sufficiently intuitive to designers, making it hard to adopt.
- *Information-theoretic.* This approach (Li 2010, 2011) also reframes the problem and solves it. Here, the problem is reframed to "maximize density of bits". While theoretically beautiful, this reframing is too complex and not sufficiently intuitive for easy designer adoption.
- *Nested MC/convolution on extracted Gaussian PDFs.* This approach (Aitken and Idgunji 2007; Abu-Rahma et al. 2008) estimates the bitcell read current PDF by running ≈ 100 MC samples, then estimating mean and standard deviation and assuming a Gaussian distribution. It estimates the sense amp PDF in a similar fashion. Then, it combines bitcell and sense amp PDFs by either nested MC sampling on the performance PDFs (Abu-Rahma Abu-Rahma et al. 2008), or by Weibull statistics and convolution (Aitken and Idgunji 2007), which are mathematically equivalent when dealing with Gaussian PDFs. Naturally, the problem is the Gaussian assumption on the PDFs.
- *Nested MC on extracted arbitrary PDFs.* This approach (Zuber et al. 2010) estimates the bitcell and sense amp performance PDFs using a high-sigma technique. Then, it combines bitcell and sense amp PDFs by nested MC sampling on the performance PDFs. Compared to the previous approach, this is an improvement because it does not assume Gaussian performance PDFs. It is fast because it only requires simulation-based high-sigma analysis at the bitcell and sense amp level. Compared to other approaches, it is easy to understand and to adopt into practical design settings. However, the approach in (Zuber et al. 2010) relied on the high-sigma technique of importance sampling, which scales poorly with a realistic number of process variables.

As we have seen, technologists have dedicated significant effort to system-level SRAM analysis. *Of all the different techniques, the last one—Nested MC on extracted arbitrary PDFs—was the most promising.* Its greatest challenge was that (Zuber et al. 2010) used a high-sigma technique that scaled poorly to medium- and high-dimensional circuits. However, we can apply HSMC, which has excellent scaling properties; it just must be adapted to output full PDFs.

The next three subsections provide detail on the components for system-level statistical analysis. The first subsection elaborates on Nested MC on extracted arbitrary PDFs, the second subsection describes how to adapt HSMC to output full arbitrary PDFs, and the third section describes how to draw samples from arbitrary PDFs.

5.7.4 Statistical System Analysis Via Nested MC on Extracted PDFs

This section elaborates on the **nested MC on extracted arbitrary PDFs** approach for system-level statistical analysis.

Table 5.4 gives the flow, for the example of a simple SRAM column. Each "For" line initiates each nested loop. The outer loop iterates through NMC_{COL} Monte Carlo samples of the column. Each column sample draws one value for sense amp offset voltage (SA *offset_v*), invokes the inner loop, and determines whether the overall column sample passes.

The inner loop iterates through N_{BIT} (e.g. 4K) bitcells, drawing one value for bitcell read current (*cell_i*) at each iteration, and updating the worst-case value (*wc_cell_i*).

Back in the outer loop, worst-case read current gets converted to a voltage, by assuming that current is charging the capacitance at a linear rate over time. The column's MC sample passes if the bitline drive voltage (*bit_v*) is large enough to be detected by the sense amp, i.e. if *bit_v* > *offset_v*. The column yield is simply the proportion of column MC samples that pass. The system yield is simply *column_yield*^{NCOL} where N_{COL} is the number of columns in the memory array (assuming no redundancy).

Table 5.4 provided an example of the approach for analyzing a simple SRAM column. The core of the approach is to mix PDFs that were extracted with a high-sigma analysis. The mixing happens via drawing MC samples from the PDFs, and applying simple logic and loops.

We can generalize Table 5.4 to broader SRAM space, to broader memory space, and to more general circuit space simply by changing the logic and the loops. Here are some example variants, in increasing order of generality:

- *Sigma vs. timing.* Edit Table 5.4 with a new loop for timing spec, that wraps the outer loop. This makes it easy to find a table of system yield versus timing spec.
- *Different interactions among components.* The equation relating bitcell read current to sense amp offset voltage is shown as *bit_v* = *wc_cell_i* * *T/C*. This assumes that bitline voltage increases linearly with time, and is inversely proportional to capacitance. But other equations could readily be incorporated.

Table 5.4 Nested MC on extracted PDFs, on example of SRAM column analysis

Input: timing spec T, bitline capacitance C, number of bitcells in column N_{BIT}, number of column MC samples NMC_{COL}
Output: $yield_{column}$

Extract PDF for bitcell *cell_i*
Extract PDF for SA *offset_v*

$N_{pass} = 0$
For $i = 1, 2, ..., NMC_{COL}$
    offset_v = abs(**Draw a sample from PDF for SA** *offset_v*)
    wc_cell_i = Inf   #worst-case *cell_i*
    For $j = 1, 2, ..., N_{BIT}$
      cell_i = abs(**Draw a sample from PDF for bitcell** *cell_i*)
      wc_cell_i = *min(wc_cell_i, cell_i)*
    bit_v = *wc_cell_i * T / C*
    column_pass = *(bit_v > offset_v)*
    if *column_pass*, increment N_{pass}
$yield_{column}$ = N_{pass} / NMC_{COL}

- *Row redundancy.* On a given column sample, track the number of bitcells N_{BIT_FAIL} that fail to have large enough read current to pass offset_v. Then, $column\_pass = (N_{BIT\_FAIL} \leq N_{RED})$ where N_{RED} is the degree of redundancy, i.e. the number of bitcells that are allowed to fail before the column fails.
- *Column redundancy.* Rather than $system\_yield = column\_yield^{NCOL}$, we allow some columns to fail before having yield loss, in a similar fashion to row redundancy.
- *Other SRAM architectures.* Simply change the logic and loops to incorporate multiplexers, error correction, delay chains, etc.
- *DRAM.* We can actually use the flow in Table 5.4 directly, by simply setting $C = 1$ and $T = 1$, so that the bitline voltage will be passed directly to compare with the sense amp offset voltage.
- *Big digital circuits.* We can extract PDFs of performance from digital standard cells, then propagate them through a large digital circuit hierarchy by applying logic and loops.

5.7.5 Full PDF Extraction via High-Sigma Monte Carlo

The last subsections provided the *motivation* to extract a full performance PDF, from −6 sigma to +6 sigma. This subsection describes *how* to do so in a fast, accurate, scalable, and trustworthy fashion.

So far, we have only applied HSMC for finding tails of distributions. It turns out we can alter HSMC to find the whole performance PDF, in a straightforward fashion. The general idea is to run HSMC at *different* numbers of generated MC samples, and *stitch together* the results into one overall PDF.

We elaborate on the approach with the aid of Fig. 5.37, which extracts the PDF for bitcell read current.

In this bitcell example, HSMC was run several times, and MC was run once. Each run had different settings, as follows:

- HSMC, max-first *cell_i*, with N_{gen} (number of generated MC samples) = 10M. Its 10 worst-case points were kept. The points are at the top right of Fig. 5.37, having sigma (quantile) ≈ 4.5 and *cell_i* ≈ 0.000019.
- HSMC, max-first *cell_i*, with N_{gen} = 1M. Its 10 worst-case points had sigma ≈ 4.0 and *cell_i* ≈ 0.0000175.
- HSMC, max-first *cell_i*, with N_{gen} = 100K.
- HSMC, max-first *cell_i*, with N_{gen} = 10K.
- MC, with $N_{gen} = N_{sim}$ = 1K. These generate the points from -2 sigma to $+2$ sigma.
- HSMC, min-first *cell_i*, with N_{gen} = 10K.
- HSMC, min-first *cell_i*, with N_{gen} = 100K.
- HSMC, min-first *cell_i*, with N_{gen} = 1 M.
- HSMC, min-first *cell_i*, with N_{gen} = 50 M.
- HSMC, min-first *cell_i*, with N_{gen} = 10G. This returned the points at the bottom left of Fig. 5.37.

The HSMC runs generated data for the 3.5–6 sigma range, and MC generated data for the 0–2 sigma range (in positive and negative sigma space).

The final step is to stitch the points together into a PDF using a 1-dimensional regression model that maps NQ to *cell_i*. The regression model can take any form, though to be mathematically correct it must be monotonically increasing. In our example, we use a piecewise linear (PWL) curve.

The outcome of full PDF extraction, for this example, is a PDF of read current (*cell_i*).

Figure 5.38 shows the outcome of extracting the PDF for sense amp offset voltage. Now that we have a full PDF for a bitcell (Fig. 5.37) and for a sense amp (Fig. 5.38), we can mix them to get statistical system-level measures for an SRAM column, as Sect. 5.7.4 described.

We have described a version of PDF extraction that includes several individual runs of HSMC, plus one individual run of MC. Clearly, these combine into a single run. Further efficiencies are gained via reuse of simulation data. From the user perspective, full PDF extraction looks simply like a special run of HSMC.

5.7.6 Drawing Samples from Arbitrary PDFs

This section describes how to draw a sample performance value v from an arbitrary nonlinear PDF, also known as random variate generation. The arbitrary nonlinear

Fig. 5.37 Full PDF extracted by HSMC, for a bitcell read current

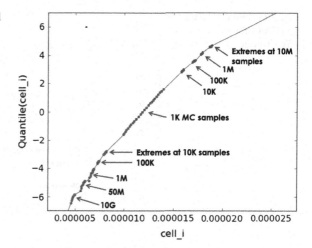

PDF is stored as a 1-dimensional regression model mapping normal quantile (NQ) values to a performance output value.

To draw a performance value from the distribution:

1. Draw a sample x from a uniform(0,1) distribution.
2. Convert the sample value x to an NQ value using the inverse cumulative distribution function (CDF) of the normal.
3. Use the PWL curve to map the NQ value to the final performance value v.

5.7.7 Example Analysis of SRAM Array

This section applies the building blocks of the previous sections—nested MC on PDFs, PDF extraction, and random variate generation—to the analysis of a simple 256 Mb SRAM array.

We followed the steps outlined in Table 5.4. First, we extracted a full PDF for the bitcell read current, using HSMC in full-PDF extraction mode, giving the results shown in Fig. 5.37. Then, we extracted a full PDF for the sense amp offset, using HSMC, to get the results shown in Fig. 5.38. Finally, we performed the rest of the analysis according to Table 5.4, as a script. We set $N_{COL} = 64K$ columns (sense amps), $N_{BIT} = 4K$ bitcells per column (per sense amp), and bitline capacitance to $C = 51.75f$. We swept across three timing specs $T = \{300, 400, 500 \text{ ps}\}$.

Table 5.5 shows the results from the analysis. The results also include lower and upper confidence intervals, at 95 % statistical confidence under the binomial distribution. As the timing spec loosens, the weakest bitcells have more of a chance to charge the bitline, and the overall column and system yields go up

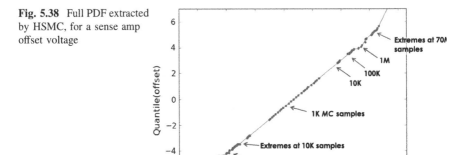

Fig. 5.38 Full PDF extracted by HSMC, for a sense amp offset voltage

accordingly. Note how even with a ≈ 3 sigma column yield for T = 300 ps, the system yield is nearly 0 % because there are so many columns. But as we loosen the timing spec to increase the column yield, eventually we can bring system yield towards ≈ 3 sigma.

5.8 HSMC Convergence/Limitations

HSMC's effectiveness and accuracy are primarily limited by its ability to generate an accurate ordering model. This section first discusses the implications of the accuracy of the ordering model on the quality of results, as well as other secondary limitations.

In the theoretical case where HSMC completely fails to order samples, HSMC degrades to MC speed and accuracy. In this case, where there is zero correlation between HSMC's ordered samples and the actual sample order, HSMC's samples are plain, unordered MC samples by definition. In this case, HSMC's error detection indicates that not all failures are found and it will continue running samples until asked to stop, producing a set of MC results, which may still be useful, though certainly not what HSMC is designed to deliver.

At the other extreme, where the ordering model is perfect, HSMC finds the actual, precise set of MC failure cases sequentially with SPICE accuracy. In this case, HSMC finds the exact tail of the failure region of the output distributions without wasting any samples, aside from those run to generate the ordering model.

The typical case is where HSMC's predicted sample ordering model has some amount of correlation, but it is not perfect. In this case, the order of predicted samples in the tails of the distributions will be imperfect. The tail of the distribution will be found, though not with optimal efficiency due to some error in the order. One essential attribute of HSMC is that, since it is running its predicted

Table 5.5 Results from example analysis of SRAM array

Probe time T (ps)	Column yield (%)	[Column lower, upper] (%)	System yield (%)	[System lower, upper] (%)
300.0	99.87	[99.868, 99.873]	1.1e-34	[2.6e-35, 4.5e-34]
400.0	99.99754	[99.99721, 99.99783]	20.712	[16.799, 24.917]
500.0	99.99999	[99.9999434, 99.9999982]	99.362	[96.439, 99.887]

samples using SPICE and getting perfect accuracy, the amount of error is discovered at runtime. Therefore, a poor correlation model implies that more simulations will be required to find all of the failures, and not that the accuracy of the result will be poor. Since a good high-sigma design with a well-chosen number of samples will have a small number of failures (e.g. <100), some inaccuracy still allows HSMC to complete verification of high-sigma designs in hundreds to thousands of simulations.

Other notable limitations of HSMC are as follows:

Number of MC samples: Since HSMC is generating and sorting real MC samples, the overhead becomes significant as the number of samples becomes large. At the time of writing, the overhead is insignificant for 100 million or fewer samples. The overhead begins to become a contributor to overall runtime past that. However, HSMC leverages parallel processing not only for simulation, but also for generating and sorting the samples. On modern machine(s) with 10 or more cores total and ≈ 60 process variables, total runtime even for 5 billion samples remains below 20 min. With a handful of process variables, runtime drops below 5 min.

Number of process variables: For reasons similar to the number of MC samples, the number of process variables is significant because each sample must generate a random number for each process variable, and each process variable is considered when sorting. Memory consumption also becomes a factor as the number of process variables increases. We have found that HSMC works well for hundreds of process variables, and recommend using it on designs with <1,000 process variables. We have observed industrial designers applying HSMC to problems with >10,000 variables, where they are willing to wait the additional time for simulation.

Not finding all failures: In some cases, HSMC may detect that it has found all failure cases, but there is still the possibility that additional failure cases are within those samples; it cannot be known for sure whether this has occurred. Fortunately, HSMC provides transparency: missed failures are more likely when the ordering model is poor, as shown by the noise in the output value vs. sample curve. For example, Fig. 5.22 is a good curve, and Fig. 5.24 is a poor curve. Even in a case where HSMC finds only half of the true failures, it nonetheless provides a reasonable indication of the distribution tail's behavior. Yield prediction is reasonable; for example, reporting 20 failures per billion versus the actual 40 per billion translates to a sigma difference of just 0.1 or so, which is still useful and sufficiently accurate for many practical purposes. In addition, HSMC has a feature that identifies "non-conforming points": sampled process points that are out-of-sync with the ordering model's prediction.

In summary, HSMC's limitations are noteworthy, but do not preclude its use with production designs.

5.9 HSMC: Discussion

This chapter presented HSMC's method, as well as demonstrated its behavior through several real circuit examples. We now examine HSMC in terms of set of qualities required for a high-sigma verification technology outlined in the introduction.

Fast: HSMC typically consumes <5,000 simulations, and often <1,000. These simulations parallelize on a cluster nearly as well as regular MC samples. The overhead required for HSMC depends on number of samples used and number of process variables, and is typically <20 min for up to 5 billion samples. To speed up the design loop further, high-sigma corners found using HSMC can be used to design against iteratively. These high-sigma corners provide accurate information about the behavior of the evolving design at the extreme tails with just a few simulations, which in turn reduces design iterations and removes the need to over-margin.

Accurate: HSMC works by leveraging the same trusted technologies used for lower-sigma verification, MC sampling and SPICE simulation. By generating a large number of real MC samples, then simulating only the samples out at the extreme tails, HSMC produces MC and SPICE accurate information at the high-sigma region of interest in the distribution. By revealing any inaccuracy at runtime by comparing the predicted order with the actual order, HSMC's predictions are reliable because it is transparent when there is inaccuracy in the sorting order.

Scalable: HSMC works for tens, hundreds and even thousands of variables. This is easily large enough to work with production designs such as bitcells, sense amps, and digital standard cells, on accurate process variation models. HSMC can verify out to true 6 sigma with MC and SPICE accuracy, which is a reasonable practical limit for high-sigma designs.

Verifiable: Since HSMC's samples are all MC samples from the true output distribution, the technology can be verified comprehensively against MC samples if the same random seed is used to generate the samples. This is useful for verifying the technology against smaller numbers of samples. Moreover, the output vs. sample convergence curve provides the key information for the designer to judge whether HSMC results has successfully found the output distribution tail yet.

In summary, to our knowledge, HSMC is the only method for verifying high-sigma designs that is simultaneously fast, accurate, scalable, and verifiable. Its availability as a CAD software product makes this effective technique easy to adopt and apply in an industrial design environment.

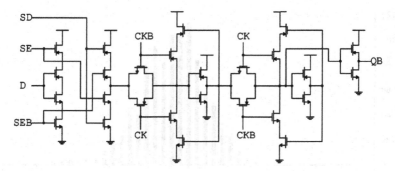

Fig. 5.39 D flip-flop schematic

5.10 Design Examples

This section presents three design examples based on the production use of HSMC by industrial circuit designers.

5.10.1 Flip-Flop Setup Time

Process variation can have a significant impact on flip-flop setup and hold time, especially at low supply voltages. Because of the important role that flip-flops serve in many digital designs, combined with the large numbers of them used in modern Systems-on-Chips (SoCs), it is important to account for process variation when designing and characterizing them in nanometer technologies (Bai et al. 2012).

In this example, a D flip-flop for a standard cell library is analyzed to determine setup time under variation conditions. This example focuses on setup time, but the procedure for other measurements (e.g. hold time) is essentially identical. Figure 5.39 shows the flip-flop schematic. Analysis is performed using Solido Variation Designer (Solido Design Automation Inc. 2012). Simulations are performed with the Synopsys® HSPICE® circuit simulator (Synopsys Inc. 2012).

Section 5.3 described many possible approaches for high-sigma analysis. Here, MC analysis with linear extrapolation and High-Sigma Monte Carlo (HSMC) are considered.

For the first step in the comparison, MC analysis is run with 1,000 samples. Figures 5.40 and 5.41 show the resulting distribution of setup time. Figure 5.40 shows the distribution in discrete PDF form (a histogram), and Fig. 5.41 shows the distribution in normal quantile (NQ) form.

From the approximate bell shape of the PDF in Fig. 5.40, and the linearity of the NQ form in Fig. 5.41, the setup time distribution appears to be approximately Gaussian. MC extrapolation predicts a 5-sigma setup time of 188.9 ps (this is where the linear extrapolation from the sampled data in the NQ plot reaches 5 sigma).

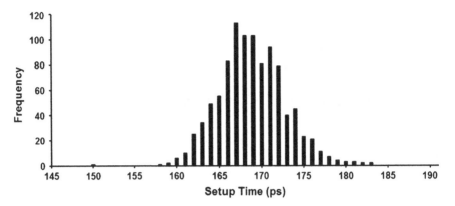

Fig. 5.40 Distribution of flip-flop setup time, in histogram form. Data is from a Monte Carlo run of 1,000 samples

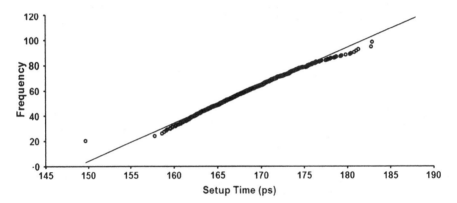

Fig. 5.41 Distribution of flip-flop setup time, in NQ form. Data is from a Monte Carlo run of 1,000 samples

HSMC is then run on the flip-flop, using 50M generated samples to accurately capture the setup time at 5 sigma. Figure 5.42 plots the tail region of the distribution found using HSMC. Figure 5.43 shows the tail region from HSMC plotted together with the 1,000 Monte Carlo samples.

From Fig. 5.43, it is clear that MC extrapolation significantly over-predicts the sigma level of the design. In fact, HSMC determines that the true sigma level of the design at 188.9 ps is actually 3.54, which is much lower than the 5 sigma predicted by MC extrapolation.

Table 5.6 shows the correct 4, 5, and 6 sigma setup times for this flip-flop, along with the setup time predicted by MC extrapolation, and the actual sigma level that *would* have resulted if MC extrapolation had been used to predict the setup time.

Clearly, MC extrapolation aligns poorly with the actual samples in the tail of the setup time distribution. This is because the distribution is not Gaussian, as was

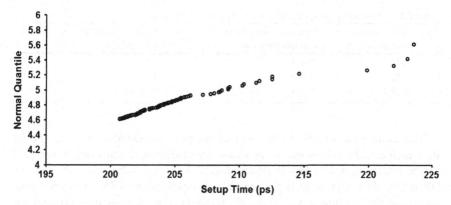

Fig. 5.42 Tail of distribution of flip-flop setup time, in NQ form. Data is from an HSMC run with 50M generated MC samples. The 100 most extreme samples are shown

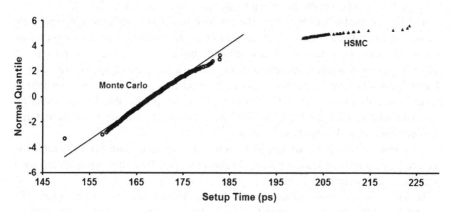

Fig. 5.43 Tail of distribution of flip-flop setup time, in NQ form. Data is from both (*1*) MC with 1000 samples, and (*2*) HSMC with 50M generated samples and the 100 most extreme samples shown

assumed when extrapolation was performed. HSMC does not make assumptions about the shape of the distribution, and therefore is able to establish the correct sigma level for this design.

5.10.2 High-Sigma DRAM Design

DRAM requires high yield across millions or billions of bitcells, and employs a write/read/restore operation that is particularly sensitive to process variability. Therefore, DRAM is an important application area for high-sigma design.

Table 5.6 Sigma and corresponding setup times for the flip-flop

Sigma	Setup time (HSMC) (ps)	Predicted setup time (MC extrapolation) (ps)	Sigma level if MC extrapolation were used
4	196.7	184.9	3.34
5	212.1	188.9	3.54
6	232.7	193.0	3.79

This example describes the high-sigma analysis of a DRAM bit slice. The bit slice includes the following components: bitcells/bitlines, sense amplifier, precharge circuit, and data bus interface. Figure 5.44 shows the schematic for the DRAM bit slice. The analysis performed in this example uses 256 cells per bitline and a precharge voltage $V_{pre} = V_{dd}/2$; however, the analysis is equivalent for other bitline lengths or for $V_{pre} = V_{dd}$. Analysis is performed using Solido Variation Designer (Solido Design Automation Inc. 2012). Simulations are performed with the Synopsys® HSPICE® circuit simulator (Synopsys Inc. 2012).

The design goal is to verify that the bit slice functions correctly at 6 sigma, by measuring the bitline voltage difference after a write/read/restore operation under statistical variation conditions. Because of the binary nature of this operation (i.e. either a 1 or 0 is written successfully or not), the bitline voltage distribution is discontinuous between a read success and a read failure. This means that failure occurs suddenly in the output distribution, so it is not predictable by extrapolating the distribution or by modeling a continuous-valued relationship between statistical parameters and output performance.

To properly verify the design to 6 sigma, a very large number of samples is required—generally at least 5 billion. In this example, 10 billion samples are used to achieve a thorough verification.

HSMC analysis is performed on this design. The analysis is set up to run 10G (10 billion) MC samples, using adaptive initial sampling (Sect. 5.6.3) to find an initial set of reasonable failures, followed by 10,000 ordered simulations adaptively targeting failures. By adaptively sampling the design, HSMC is able to find several failures that occur out of the 10 billion MC samples, and determine that the sigma level of the design is 5.955 sigma. This is just below the target of 6 sigma; some design iteration could be performed using one or more of the failing samples in order to bring the design above 6 sigma.

Figure 5.45 shows the NQ plot generated from the HSMC run. It clearly shows the discontinuous behavior described above. An output voltage of 1.2 V corresponds to a successful read operation, while an output voltage of −1.2 V corresponds to a failed read. HSMC successfully captures the failing samples even though the failure mode is binary.

It is sometimes possible to define intermediate measurements that behave more continuously, in an attempt to avoid binary output conditions. For example, in this design, it may be possible to measure the pre-sense bitline voltage levels and the sense amplifier offset independently. Doing so can often improve the efficiency of HSMC. However, in many cases it is not possible to ensure that failures in these

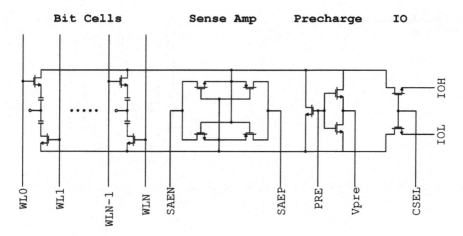

Fig. 5.44 DRAM bit slice schematic

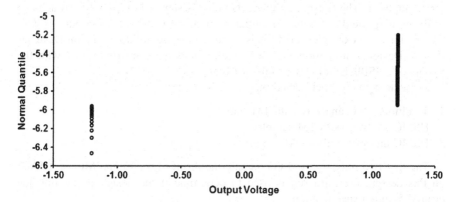

Fig. 5.45 Tail of distribution of DRAM bit slice output voltage, in NQ form, from HSMC with 10G generated MC samples. The 100 most extreme samples are shown

intermediate measurements are directly correlated with the true failure rate of the design. Therefore, it is important to be able to properly simulate the true failure mode of the design even when the key performance metrics are binary. The adaptive initial sampling technique in HSMC (Sect. 5.6.3) makes this possible.

5.10.3 SRAM Sense Amplifier

The yield of an SRAM design is heavily dependent on the variation resilience of both its bitcells and its sense amplifiers. Sense amplifiers are highly repeated components, and they need to operate properly under variation conditions in order to avoid read failures.

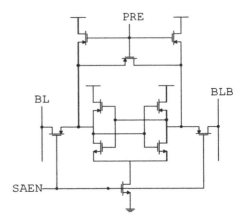

Fig. 5.46 Sense amplifier schematic

Consider the following example of how HSMC is used to analyze an SRAM sense amplifier. The sense amplifier schematic is shown in Fig. 5.46. To achieve sufficient chip yield, the sense amplifier needs to be verified to 5 sigma. As with the previous two examples, analysis is performed using Solido Variation Designer (Solido Design Automation Inc. 2012), and simulations are performed with the Synopsys® HSPICE® circuit simulator (Synopsys Inc. 2012).

Three analyses are performed on this sense amplifier:

1. Traditional MC analysis with 1M samples
2. HSMC analysis with 1M samples
3. HSMC analysis with 100M samples

The first two runs are used to compare the result of traditional MC vs. HSMC on this design; the third run is used to determine if the design is meeting the desired 5-sigma specification.

Table 5.7 shows the results of each of these three runs.

For the MC run, it takes 8 days to simulate 1M samples, and all 1M samples pass specification. For the same analysis using HSMC, only 900 simulations are required and the analysis takes only 8 min and 11 s.

1M samples is only enough to verify to approximately 4.5 sigma, so the second HSMC run is performed to verify to the target of 5 sigma. As shown in the table, when 100M samples are run, the sigma of the design is found to be 4.790, which is below the target of 5 sigma. Therefore, some changes will need to be made to this sense amplifier in order for it to reach its robustness goal. These changes are outside the scope of this example, but possible changes might include adjusting the supply voltage, adjusting the offset specification, or resizing the sense amplifier.

The 100M-sample analysis takes only 9 min and 51 s to run, in contrast to the 800 days it would have taken using traditional MC analysis. Note that this is only 1 min and 40 s longer than the high-sigma run with 1M samples. This is because, for this design, ordering and data processing time is much faster than the

Table 5.7 Sense amplifier results for traditional MC and for HSMC

	Total # of samples	# of samples simulated	Simulation savings	Predicted sigma	Predicted # of failures	Run time
Monte Carlo	1M	1M	N/A	$+\infty$	0	8 days
HSMC	1M	900	999,100	$+\infty$	0	8 m 11 s
HSMC	100M	900	999,999,100	4.790	167	9 m 51 s

simulation time, so very little additional time is required to increase significantly the number of samples used.

For readers who are new to high-sigma analysis, it may appear surprising that there are no failures in 1M samples while there are 167 failures in 100M samples. The reason for this is simply that 1M samples are not enough to reach into the tail of the distribution where the failures occur for this design. From the 100M-sample run it can be seen that the failure rate is 167/100M = 1.67e-6, or 1.67 failures per million. In this example, due to the nature of random sampling, it happened that none of those failures occurred in the 1M samples that ran. If the run were repeated with a different random seed, it is possible that one, two, or even more failures could be found by chance in a 1M-sample run. But even then, without running more samples as was done with HSMC, there's no way to know the true sigma level of the design with any certainty. HSMC makes it possible to properly determine this sigma level.

5.11 Conclusion

Semiconductor profitability hinges on high yield, competitive design performance, and rapid time to market. For the designer, this translates to the need to manage diverse variations (global and local process variations, environmental variations, etc.), and to reconcile yield with performance (power, speed, area, etc.), while under intense time pressures.

With high-sigma designs, where failures are one in a million or a billion, previous approaches to identifying failures and verifying those designs were either extremely expensive, inaccurate, or not trustworthy.

High-Sigma Monte Carlo (HSMC) is a new approach for high-sigma analysis. It generates millions or billions of MC samples, then simulates a small subset to find extreme tail values. It is fast, accurate, scalable, and verifiable. It enables rapid high-sigma design by enabling high-sigma feedback within the design loop and by making high-sigma corners available to design against. HSMC also serves as an excellent mechanism for verifying high-sigma designs, reducing verification time and improving accuracy over conventional methods. This in turn promotes the reliable development of more competitive and more profitable products that rely on high-sigma components. HSMC with full-PDF extraction enables practical, accurate statistical analysis of memory systems and other large systems.

References

Abu-Rahma MH, Chowdhury K, Wang J, Chen Z, Yoon SS, Anis M (2008) A methodology for statistical estimation of read access yield in SRAMs. In: Proceedings of design automation conference (DAC), June 2008, pp 205–210

Aitken RC, Idgunji S (2007) Worst-case design and margin for embedded SRAM. In: Proceedings of design automation and test in Europe (DATE), March 2007, pp 1289–1294

Bai X, Patel P, Zhang X (2012) A new statistical setup and hold time definition. In: Proceedings of international conference on integrated circuit design and technology (ICICDT), May 2012

Celis M, Dennis JE, Tapia RA (1985) A trust region strategy for nonlinear equality constrained optimization. In: Boggs P, Byrd R, Schnabel R (eds) numerical optimization, SIAM, Philadelphia, pp 71–82

Drennan PG, McAndrew CC (2003) Understanding MOSFET mismatch for analog design. IEEE J Solid State Circuits 38(3):450–456

Gu C, Roychowdhury J (2008) An efficient, fully nonlinear, variability-aware non-Monte-Carlo yield estimation procedure with applications to SRAM cells and ring oscillators. In: Proceedings of Asia-South Pacific design automation conference (ASP-DAC), pp 754–761

Hastie T, Tibshirani R, Friedman J (2009) The elements of statistical learning, 2nd edn. Springer, NY

Hesterberg TC (1988) Advances in importance sampling. PhD Dissertation, Statistics Department, Stanford University

Hohenbichler M, Rackwitz R (1982) First-order concepts in system reliability. Struct Saf 1(3):177–188

Hocevar DE, Lightner MR, Trick TN (1983) A study of variance reduction techniques for estimating circuit yields. IEEE Trans Comput Aided Des Integr Circ Syst 2(3):180–192

Kanj R, Joshi RV, Nassif SR (2006) Mixture importance sampling and its application to the analysis of SRAM designs in the presence of rare failure events. In: Proceedings of design automation conference (DAC), June 2006, pp 69–72

Kanoria Y, Mitra S, Montanari A (2010) Statistical static timing analysis using Markov Chain Monte Carlo. In: Proceedings of design automation and test in Europe (DATE), March 2010

Li X (2010) Maximum-information storage system: concept, implementation and application. In: Proceedings of international conference on computer-aided design (ICCAD), pp 39–46

Li X (2011) Rethinking memory redundancy: optimal bit cell repair for maximum-information storage. In: Proceedings of design automation conference (DAC), pp 316–321

McConaghy T (2011) High-dimensional statistical modeling and analysis of custom integrated circuits. In: Proceedings of custom integrated circuits conference (CICC), September 2011, pp 1–8 (invited paper)

Metropolis N, Rosenbluth AW, Rosenbluth MN, Teller E (1953) Equations of state calculations by fast computing machines. J Chem Phys 21(6):1087–1092

Niederreiter H (1992) Random number generation and quasi-Monte Carlo methods. Society for Industrial and Applied Mathematics (SIAM), Philadelphia

Qazi M, Tikekar M, Dolecek L, Shah D, Chandrakasan A (2010) Loop flattening and spherical sampling: highly efficient model reduction techniques for SRAM yield analysis. In: Proceedings of design automation and test in Europe (DATE), March 2010

Schenkel F et al (2001) Mismatch analysis and direct yield optimization by spec-wise linearization and feasibility-guided search. In: Proceedings of design automation conference (DAC), pp 858–863

Singhee A, Rutenbar RA (2009) Statistical blockade: very fast statistical simulation and modeling of rare circuit events, and its application to memory design. IEEE Trans Comput Aided Des 28(8):1176–1189

Solido Design Automation Inc. (2012) Variation designer. http://www.solidodesign.com

Synopsys Inc. (2012) Synopsys® HSPICE®. http://www.synopsys.com

Wang J, Yaldiz S, Li X, Pileggi L (2009) SRAM parametric failure analysis. In: Proceedings of design automation conference (DAC), June 2009

Wilson EB (1927) Probable inference, the law of succession, and statistical inference. J Am Statist Assoc 22:209–212

Zuber P, Dobrovolný P, Miranda M (2010) A holistic approach for statistical SRAM analysis. In: Proceedings of design automation conference (DAC), June 2010, pp 717–722

Chapter 6
Variation-Aware Design

Manual Sizing, Automated Sizing, and an Integrated Approach

Abstract Previous chapters focused on the analysis of a *fixed* design, and corner extraction and verification in particular. Complementary to those chapters, this chapter focuses on *changing* the design. Specifically, it explores three different methodologies for changing device sizings, given accurate corners. First, it explores manual sizing and the advantages and disadvantages of a manual approach. Second, it explores automated sizing and its associated advantages and disadvantages. Finally, this chapter introduces a new idea that integrates manual and automated design techniques. This integrated approach incorporates the benefits of both manual and automated design, providing fast and thorough design exploration while allowing the designer to maintain full control over the design and providing more insight than ever.

6.1 Introduction

Design is the central part of value creation in integrated circuits (ICs). It is through design where circuits with wholly new functionality are created, or more commonly, where existing functionality is enhanced through faster, better, and smaller circuits. In custom ICs, design is even more nuanced because the tradeoff among different speed, performance, and area is a continuum rather than a small fixed set of choices.

It is in design where creativity, insight, and experience can all make enormous contributions to the quality of the final circuit. These factors may lead to better *choices* of circuit topologies and architectures, and *new* topology designs, which means better circuits. These factors also affect the choice of device sizes, which have enormous impact on the circuits.

T. McConaghy et al., *Variation-Aware Design of Custom Integrated Circuits:*
A Hands-on Field Guide, DOI: 10.1007/978-1-4614-2269-3_6,
© Springer Science+Business Media New York 2013

Fig. 6.1 This chapter
focuses on design, whereas
other chapters focused on
verification and corner
extraction steps

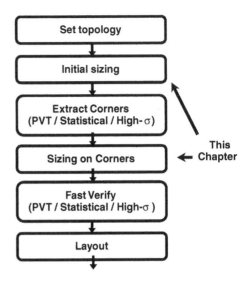

Variation effects can get in the way of the core task of design, of exploring topologies and device sizings. Traditionally, designers have had poor visibility into accurate effects of variation during their design cycle, especially if they wanted design iterations to be fast. However, since variation has a big impact on performance and yield, ignoring variation in the design loop hurts the final design. The hard-won design effort to make a great circuit may be for naught. For a truly good design, the effects of design choices (the controllable variables) and the effects of variation (uncontrollable variables) must *both* be considered during the design phase.

As first introduced earlier in this book, there *is* a way to achieve rapid design iterations while accurately capturing variation effects. It does not require a radically new methodology. It simply requires *accurate* corners; corners that represent the bounds of performance subject to PVT or to statistical variation. These corners enable rapid design iterations that are accurate to variation. Other chapters of this book described how those corners could be extracted and how the final design could be verified in a trustworthy fashion. Those tools are meant to be as pushbutton as possible: get the corners with minimal fuss, so that the designer can get on with actually *designing*. Once design is done, verify the circuit, but also with minimal fuss. The whole point of variation-aware analysis is to do just enough to be variation-aware, then get out of the way so that the core design process can happen.

Figure 6.1 illustrates the corner-driven variation-aware flow and shows how this chapter fits in. Whereas the other chapters of this book explored corner extraction and verification, this chapter explores design with emphasis on device sizing. The design step occurs in two places: against corners, but also in initial sizing where there is simply a single nominal corner.

Traditionally, sizing of custom ICs has been a manual process. Automated sizing has frequently been proposed as a means to accelerate the design cycle, but adoption has been sporadic for various reasons. We believe there is third way that builds on manual design, yet brings the benefits promised by automated sizing. This is done by integrating manual and automated design.

It is useful to explore all three techniques in the context of a variation-aware design flow. Accordingly, this chapter has three parts:

- First, it discusses **manual approaches** to design, which remain as valid and useful as ever because they help the designer retain insight into the design.
- Second, this chapter discusses **automated approaches** with a focus on automated sizing.
- Finally, this chapter introduces an **integrated approach**, which, like automated design, has fast and thorough design exploration, but like manual design, allows the designer to maintain full control over the design and gain more insight than ever.

Regardless of which of these three design approaches is used, the designer can achieve high-performing, high-yielding designs, thanks to the corner-driven methodology.

6.2 Manual Design

6.2.1 Introduction

Manual custom IC design includes two parts: topology selection and design, and device sizing. Topologies are selected based on experience, design knowledge, and analysis. Topologies are designed or modified based on the latter plus a healthy dose of creativity. Creativity abounds. It is innovation in custom circuit topologies that drives the whole field of custom IC design. This is perhaps best represented by publications in the field's conferences and journals such as the International Solid-State Circuits Conference (ISSCC), the Custom Integrated Circuits Conference (CICC), and the IEEE Journal of Solid-State Circuits (JSSC).

Sizing is another central aspect of manual design. There are two complementary tools in manual design: hand-based analytical methods, and SPICE-based methods.

Hand-based analysis and design: Designers often use hand-based analytical methods to find initial device sizings. Structured manual flows exist, in which the designer incrementally reduces the number of free design variables by applying reasonable rules of thumb and adding constraints (Sansen 2006). For example, in designing an amplifier, the first step is to set all lengths to the process minimum. Then, given an overall power budget and supply voltage, the next step is to allocate currents to each current branch. Then, the designer allocates an overdrive voltage for each device channel. Then, the designer adds a performance constraint, reducing the free variable count by one. The designer incrementally adds more constraints until no free variables remain. Finally, the designer computes device widths using first-order analytical equations.

SPICE-based analysis and design: Traditionally, designers have used hand-based analytical methods to choose the sizings, then verified the design via a small set of SPICE simulations. SPICE was not used for tuning. However, on modern process geometries, the first-order equations of hand-based methods are less accurate, especially when there are variation effects. Since SPICE can readily model the second-order and variation effects, designers have turned to using SPICE *within* the design loop. They invoke a series of SPICE runs to gain design insight, for example, using sensitivity analysis, then use the results to decide what design variables to change, and by how much.

6.2.2 SPICE-Based Analysis Techniques

A variety of SPICE-based analyses and tools are available to provide insight to designers. Some examples include operating point analysis, waveform analysis, eye diagrams, histograms, NQ plots, and sensitivity analysis. New techniques continue to be developed, such as noise analysis of phase-locked loops (Mehrotra 2000) and nonlinear symbolic analysis (McConaghy and Gielen 2009a).

The most common SPICE-based analyses for measuring the **effect of design variables** are sensitivity analysis, sweep analysis, and combinational sweep analysis. These analyses have value because they are directly actionable: they point to which design variables can be changed and by how much, which of course has a direct effect on the circuit's performance and yield. Each of these three analyses reports increasingly more information, but at increasingly higher simulation cost. Figure 6.2 illustrates the three analyses on two design variables $W1$ and $W2$. Each dot is another analysis that is simulated. We now describe each analysis in more detail.

Sensitivity analysis (Fig. 6.2a) perturbs a design variable by a small amount from a reference design, while all other design variables are held fixed. SPICE is run against corners at that new design point. This perturb-and-simulate process is repeated for each design variable of interest. This analysis reports the sensitivity of each output to each design variable in the local region near the reference design point. This analysis can be viewed as a local linear approximation of the mapping from design variables to performance. There are actually two variants: perturbing in just one direction ($+\Delta$), or in both directions ($+\Delta$, $-\Delta$). If n_{var} is the number of design variables, then sensitivity analysis requires $(n_{var} + 1)$ simulations per corner for one direction, or $(2 \times n_{var} + 1)$ per corner for both directions.

Sweep analysis (Fig. 6.2b) sweeps a design variable across several values, while all other design variables are held fixed using the value of the reference design. SPICE is run on each swept design value. This process is repeated for each design

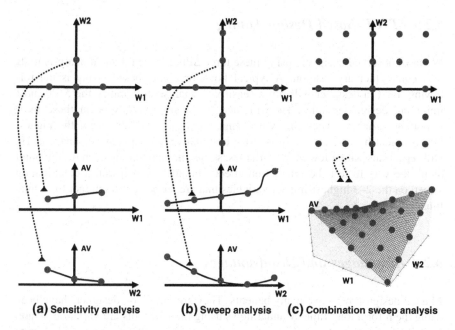

Fig. 6.2 The most common SPICE-based design analyses: **a** sensitivity analysis, **b** sweep analysis, and **c** combination sweep analysis. Each *top plot* shows the samples taken in design variable space *W1* and *W2*. Each *bottom plot* shows the mapping from each design point to a corresponding output performance value *AV*

variable of interest. This analysis reports the local sensitivity to each design variable, as well as the effect of each variable across a broad range, but for only one variable at a time. It can be viewed as a broader model of circuit performance, with the caveat that changes to >1 design variable will be assuming weak interactions among the variables. If n_{val} is the number of swept values per variable (including the center value), then sweep analysis requires $(2 \times n_{var} \times (n_{val}-1) + 1)$ simulations per corner.

Combination sweep analysis (Fig. 6.2c) considers all combinations (cross-product) of all variables' values. Each combination point is simulated. This analysis reports not only the local sensitivity to each design variable and the effect of design variable across a broad range, it also considers the non-additive interactions among the design variables with respect to their influence on output performances (e.g. $W1 \times W2$, or $W1/L1$). It accounts for all design variable interactions for the whole range that is sampled. Combination sweep analysis requires $(n_{val})^{var}$ simulations per corner. It gives the most information, but because it depends exponentially on the number of design variables, it gets extremely expensive even at moderate numbers of variables. For example, with 10 values per variable and one corner, then 4 variables would require 10,000 simulations, and 10 variables would require 10 billion.

6.2.3 SPICE-Based Design Tuning

We have observed designers using these three SPICE-based manual analysis tools in a complementary fashion. A typical flow for corner-based sizing is the following. First, a designer will run a sensitivity analysis to identify the 5–20 most important design variables. He will then run a sweep analysis on these most important variables. Then, he will change the design variable with the biggest improvement from sweep analysis to the best value returned in sweep analysis. He will repeat this with a few more variables, sometimes simulating each design point along the way to test the interaction effects. Finally, he may run a combination sweep on the 2–3 highest-impact variables that are suspected to interact for a final tuning.

6.2.4 Advantages and Disadvantages

Manual design flows have many benefits. They are familiar—designers have been using them for a long time. Manual flows using modern tools are variation-aware: by using accurate corners, designers may choose sizings that perform well, despite significant performance variations caused by PVT or statistical process effects. Put another way, one does not need to resort to automation just because there is variation. Topology changes are easy and inexpensive: designers can modify an architecture and keep using the same corners. The corner may lose some accuracy, but that is fine because the final verification step will catch any inaccuracies. Finally and most importantly, manual design flows allow the designer to maintain **insight and control**: the designer has intimate understanding of how the choice of topology and sizings affects performance; if there is ever an issue, he knows where and how to take action. Building insight is typically part of the flow itself. For example, hand-based analysis includes development of equations relating design variables to performance, and SPICE-based analysis sweeps find the mapping from design variables to performance.

Manual flows have some disadvantages. Most obviously, they require designer time, which is always at a premium. Circuit quality will suffer if a designer cannot spend sufficient time on design to explore the design space. Manual flows also require designer expertise; inexperienced designers might not focus their efforts in the most effective ways, and end up with suboptimal designs (but of course the very act of designing will increase the designer's level of experience). Finally, some manual design flows may result in variation-sensitive designs.

Appropriate tools can largely moderate the disadvantages of manual flows. The tools should make it easy to find accurate corners, to perform sensitivity and sweep analyses across the corners, to visualize the results conveniently, to identify the most important variables, and to act upon the insight gained.

6.3 Automated Design

6.3.1 Introduction

While manual design is familiar to designers, supports variation-aware design, promotes insight, and lets designers maintain control over their circuits, there are cases where it might take longer than the time available in order to get a high-quality circuit. An example is in porting a whole library of digital standard cells, where there is a small design team, a tight schedule, yet a large number of circuits that need to be moved to a new process. In scenarios like this, automated design might be a reasonable consideration.

Automated design of custom circuits may be at the schematic (front-end) level, layout (back-end) level, or both. Schematic-level design may involve automated topology design or automated device sizing. Automated layout design includes automatic device generation, placement, and routing. Here, we will focus on schematic-level automated design.

6.3.2 Automated Topology Design

While automated topology design is not currently a practical possibility for industrial-scale circuit design, we briefly review it in order to raise awareness and for the sake of thoroughness.

Automated topology design may involve topology selection or topology synthesis. Automated topology selection searches a set of trusted topologies to find a topology most suitable to the design problem at hand. Topology synthesis automatically chooses a set of devices and their connections to realize the behavior most suitable to the problem at hand.

There has been a steady stream of research on automated topology selection and synthesis over the years; (McConaghy et al. 2009) has a comprehensive survey, and more recent work appears in (Meissner et al. 2012). Topology selection across a sufficiently large set of topologies will start to feel like synthesis, such as (Palmers et al. 2009), which searched a database of 101,904 possible topologies. Research on topology *analysis* is now emerging, such as (Ferent and Doboli 2011).

Automated topology design tools are not commercially available yet, with the exception of (Magma Design Automation 2012), which does topology selection across a small set of pre-characterized topologies. Its roots trace back to (Hershenson et al. 2001).

6.3.3 Automated Sizing: Problem Types

In automated sizing, an optimizer automatically determines a design point (set of device sizes) that best meets the circuit performance objectives and constraints. Let us assume here that the designer is interested in using an optimizer. If he has access to an optimizer, that optimizer will likely be designed for a certain set of problem types. Since it will be instructive for the designer to be aware of the possible problem types, we give the following dimensions for classifying problem types:

- *Constraint-satisfying versus single-objective versus multi-objective*: A constraint-satisfying optimizer aims to find a design point that meets all constraints. A single-objective optimizer aims for a design point that maximizes or minimizes a given objective (e.g. minimize power), while meeting design constraints (e.g. AV $\geq$ 70 dB). A multi-objective optimizer aims for a set of design points that collectively trade off a set of objectives (e.g. power versus AV), subject to a set of constraints.
- *Local versus global*: Local optimizers "run up a hill", but get stuck in local optima. In contrast, global optimizers aim to find the global optimum by avoiding local optima. Ideally, a global optimizer is *globally reliable*—the user should not need to worry about whether or not the algorithm is stuck at a local optimum.
- *Nominal versus variation-aware*: Nominal optimizers only measure performance under nominal and typical conditions, whereas variation-aware optimizers consider PVT and/or statistical variations. Variation-aware optimizers have many variants, including: optimize against a set of corners, maximize yield, and maximize the distance from the design point to the closest spec boundary ("design centering").
- *Equation-based versus SPICE-based*: Equation-based optimizers measure performance using symbolic models which were hand-generated or automatically generated. SPICE-based optimizers measure performance using SPICE in the sizing loop. Some optimizers may generate equation-based models or response surface models during optimization, but since those models are based on feedback from SPICE, the optimizers are nonetheless SPICE-based.

Many optimization algorithms have been proposed for different combinations of problem dimensions above. For example, (Gielen and Sansen 1991) is single-objective, global, nominal, and equation-based. (Krasnicki et al. 1999) is single-objective, global, nominal, and SPICE-based. (McConaghy and Gielen 2009b) is single-objective, global, variation-aware, and SPICE-based.

6.3.4 Automated Sizing: Optimizer Criteria

For any given type of optimization problem, there is a huge variety of algorithm options, such as derivative-based Newton methods (e.g. Nocedal and Wright 1999), derivative-free pattern search (e.g. Kolda et al. 2003), simulated annealing (Kirkpatrick et al. 1983), evolutionary algorithms (e.g. Hansen and Ostermeier 2001), convexified search (Boyd and Vandenberghe 2004), and response surface-based/model-building optimization.

Accordingly, optimizers may have dramatically different behavior and performance. The quality of the optimizer can be rated according to the following:

- *Quality versus runtime*: A good optimizer will deliver high quality circuits in minimal runtime. Sometimes, a longer runtime is tolerable if it produces even better quality circuits. While minimizing simulations matters, runtime is more relevant because it accounts for parallelization across cores and machines. Quality can be higher with global (versus local), variation-aware (versus nominal), and SPICE-based (versus equation-based) problem types. For example, convex equation-based optimization may return a circuit with good performance on the convex model, but have poor performance when subsequently simulated with SPICE.

- *Flexibility*: It should be easy for designers to set up optimization on a new topology. It should be easy to choose which design variables to optimize on, and what ranges those variables should have. It should be easy to set up measures of performance, area, and yield; and assign those to objectives or constraints. For example, automatic extraction of device operating constraints may be useful (Graeb et al. 2001). Or, SPICE-based approaches tend to be more flexible than equation-based approaches, because SPICE can be used off-the-shelf for a broad range of circuit types, whereas equation-based analysis is highly dependent on circuit type and likely requires setup effort for each circuit type and each new topology.

- *Scalability*: A good optimizer will scale to handle the number of design variables for the target application without becoming unreasonably expensive. Many response surface model-based optimizers are efficient, but scale poorly beyond 25–50 dimensions. Good optimizers should also be able to handle a large number of constraints, and, if doing multi-objective optimization, a large number of objectives.

Research in automated sizing stretches back several decades in the analog CAD literature (Rutenbar et al. 2002; Rutenbar et al. 2007). Starting in the late 1990s, competent commercial offerings became available. Analog Design Automation Inc. (ADA), Barcelona Design Inc., Opmaxx Inc., and Neolinear Inc. were among the first. ADA and Neolinear are now part of Synopsys Inc. and Cadence Design Systems Inc., respectively. Since that time, there have been offerings by MunEDA GmbH, Orora Inc., and many others.

The enthusiastic efforts in automated sizing by the CAD community has not translated to widespread industrial adoption by custom IC designers. Looking back, perhaps we should not be surprised: manual design has a great number of benefits, and it is hard for automated design tools to meet some of those benefits. Most importantly, doing design manually improves a designer's insight and maintains a designer's control over the design. Doing design automatically does neither.

6.3.5 Automated Sizing: Example on Digital Standard Cells

As discussed, there are some cases where automated sizing (versus manual) may be palatable to designers. One example is in process migration of digital standard cells, where hundreds of standard cells must be updated to the new process. We illustrate the procedure to first classify the problem into the appropriate type, consider optimizer criteria for the problem, and finally design an optimization algorithm that does well according to the problem type and optimizer criteria.

Problem Type: For digital standard cells, a reasonable choice of optimization problem type is: single-objective, global, variation-aware, and SPICE-based. It is single-objective because most performance aims can be set as constraints. It is global, variation-aware, and SPICE-based because the extra computational effort is worth the improved design quality; the cells are used numerous times throughout the design, and their quality can have a big impact on the overall circuit's power, speed, and area. Furthermore, standard cells are sufficiently small such that global optimization and SPICE simulation have tractable computational requirements.

Optimizer Criteria: We want the optimizer to perform well with respect to the criteria of quality versus runtime, flexibility, and scalability. Since digital standard cells typically have fewer than 50 design variables, scalability will be less of an issue compared to other circuit types.

Optimizer Design: To make the optimizer variation-aware, let it be the variation-aware sizing step in the flow of Fig. 6.1. Therefore, it will use accurate corners that were extracted using PVT or statistical analysis.

For flexibility, we ensure that the user can readily construct an objective and constraints from different output SPICE measurements. The SPICE-based characteristic also helps flexibility.

To make the approach have good quality versus runtime, let the optimizer use adaptive response surface modeling to make maximal use of simulations taken so far when choosing promising new designs to simulate. The models will be arbitrarily nonlinear, to handle any type of circuit. To be global—to avoid local optima in the form of model blind spots—the models report the confidence in their own predictions. The optimizer will sample design regions that may be good, but not enough is known yet to be certain. In the end, the optimization will return globally-optimal designs with confidence.

Fig. 6.3 Algorithmic flow
for model-based optimization

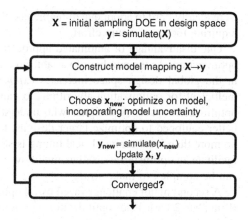

Figure 6.3 shows the exact steps of the model-based optimization.[1] The first step
of the inner loop draws a set of initial samples in design space, X, then simulates
them to get estimates of each output, e.g. power (y_{power}) and delay (y_{delay}). Then, it
constructs a model mapping $X \rightarrow y$ for each output; that model can predict its
uncertainties. Using those models, it computes a new candidate design that
combines the prediction of maximum performance, combined with some model
uncertainty. It simulates the new design, and updates the training data X, y accordingly. It will re-loop back to building new models, until it converges (e.g. no recent
improvements), at which point it will return back to the outer loop (a).

In the optimization algorithm described, the optimization approach was on a
pre-set group of corners, maximizing the worst-case performance across the corners. But it can be applied to optimize yield as well, by including corner extraction
and verification as part of the loop, as in Fig. 6.1. These combined steps may be
fully automated; simply chain together corner extraction, optimization on corners,
and verification into the same script.

6.3.6 Advantages and Disadvantages

Manual approaches are widely used, but require dedicated designer time. Automated approaches have promised to improve sizing turnaround time and help find
optimal-quality designs, yet they have not been widely adopted. Some optimizers

---

[1] Discerning readers will see that the algorithm is similar to the algorithm for Fast PVT
verification in Chap. 2. This should not be surprising: both problems aim to maximize or minimize a performance with low computational effort, and have ≤ 50 dimensions, which makes
model-building optimization tractable. The big difference between the algorithms is in the
stopping criteria. In Fast PVT verification, the optimizer has to be *really* sure that the worst-case
is found; whereas automated sizing can stop anytime, but preferably with a big improvement to
the initial design.

were clearly hard to adopt due to inefficient algorithms leading to long runtimes, or requiring too much setup effort.

One disadvantage of automated approaches is the limited scalability of global optimization. As the number of variables increases, the number of possible designs in the global design space increases *exponentially*. At some point, it becomes inefficient for optimization algorithms to gain even sparse coverage of the global optimization space. A designer who understands how the design works may be better equipped to optimize larger designs, leaving the job of the optimizer to do no more than local tuning. Local tuning is still valuable, though most tools do not facilitate an optimizer-assisted manual flow, and although potentially valuable, this can be complex to make work.

A second major challenge faced by all optimizers is having adequate constraints (Rutenbar 2006). If an optimization problem is not adequately constrained, then the optimizer may adapt towards design regions that it believes are good, but by the designer's judgment are not. For example, the optimizer may return devices with extreme aspect ratios if that constraint is not pre-constrained by the designer. Some constraints can be automatically set, such as restrictions on device operating regions (Graeb et al. 2001). But other issues are only noticed once the optimizer returns an unacceptable design; at which point the user has to add another constraint and re-run the optimization. This cycle of filling "holes in goals" (McConaghy et al. 2009) can become very tedious, and can have a serious effect on the usability of optimizers. This challenge can be somewhat mitigated with sufficient support for manual and automatic constraint generation in the optimization tool for appropriate sub-problem domains (such as digital standard cell sizing).

All optimizers share an even greater challenge: they do not afford opportunities to learn about the mapping from design variables to outputs, which compromises the designer's insight. Furthermore, the optimizer "pushes" the sized circuit to the designer, effectively taking control of the design from the designer. This loss of insight and control *really* matters, because designers need to be equipped to solve issues when things go wrong. The list of things that can go wrong is long. For example, the existing topology is inadequate and must be changed; the simulator does not converge; layout-induced parasitics heavily degrade performance; or previously-ignored variation effects like well proximity (Drennan and Kniffin 2006) or aging (Maricau and Gielen 2010) affect the circuit.

6.4 Integrating Manual and Automated Approaches

6.4.1 Introduction

So far, this chapter has focused on manual and automated approaches for automated sizing. Manual approaches are familiar to designers, support variation-aware design, and most importantly, allow designers to maintain insight and control over their circuits. Automated approaches have promised to improve sizing turnaround time and help find optimal-quality designs, but they compromise insight and control, are limited to small designs or designs with good starting points, and need good constraints.

Intuitively, what designers really need is an approach that integrates ideas from both manual and automated design, and delivers the best attributes of both techniques. Specifically, an integrated approach that is all of:

Efficient: Helps the designer to complete their design with less effort and wall clock time. In particular, it automates tedious sizing tasks, such as sizing for design porting or trial-and-error iterations with SPICE to squeeze out the last bit of performance.

Thorough: Reliably determines the optimal design, as long as the design space is appropriately limited. Considers the global space, but able to follow design hints input from the designer.

Insightful: The tool shares what it has learned about the design with the designer in ways that are understandable and familiar. It comes with clear visualizations that help to ensure that the designer not only still understands how the design works, but actually understands it better than before. In particular, the tool could improve insight into the mapping from design variables to output performances.

Controllable: The designer has full control over the design outcome. Possible changes are communicated clearly to the designer as suggestions with clear disclosure as to what would be changed and why. Design regions being sampled that are not appropriate can be halted in real time, and the designer can interactively guide the tool toward more promising improvements and design regions.

The user interface for an integrated approach is tricky, as it must be able to both present insight and design suggestions to the designer in a useful and actionable manner, as well as accept guidance from the designer efficiently. This section presents an idea for what such a user interface might look like.

6.4.2 Design Exploration User Interface

6.4.2.1 The Complexity Problem

In sizing, the designer has to consider the effects of several design variables on several output performance measures, subject to variation. Even if the information about the effect of all design points was available for free, considering these all at once is overwhelming. Let us consider if we had 200 design variables, 10 outputs, and 5 corners to capture variation effects. Which output should we focus on improving? Against which corner? And which design variables should we change to improve it? If we ignored interactions among design variables and considered just main effects, then there are $200 \times 10 \times 5$ possible combinations of (design variable, output, corner) that we might examine. If we also wanted to account for two-variable interactions among design variables, there are $(200)(200 - 1)/2 \approx 20{,}000$ interactions, and $20{,}000 \times 10 \times 5 = 1$ million possible combinations of (design variable, output, corner) that we might examine.

6.4.2.2 Achieving Tractable Complexity

Clearly, we need a way to reduce the complexity of the problem. To do so, we state the tool interface design problem as: *guide the user towards making the greatest-benefit design choices first*. We want to help the designer focus on the design variables giving greatest improvement against the outputs and corners that are causing the most problems. Put another way, we aim to reduce complexity on three fronts: design variables, outputs, and corners. We can do this as follows:

- *Reduce corner complexity* by displaying the *worst-case* output performance across corners, at any given design point.
- *Reduce output complexity* by creating a new surrogate output called "overall margin" that combines all output values according to how well they have met, or not met, specifications. Margin is >0 when specs are met, <0 when not met, and $=0$ when precisely met but not exceeded. Specifically:
  - For a given worst-case output value v, compute $margin_{output}(v) = (v - spec)/range$ for "$\geq$" constraints, and $margin_{output}(v) = (spec - v)/range$ for "$\leq$" constraints, where $range = max(max$ simulated $v, spec) - min(min$ simulated $v, spec)$
  - We now have a margin value for each output. Because all outputs now have the same units (for margin), we can compare *across outputs* to compute an overall value $margin_{overall} = min(margin_{output1}, margin_{output2}, \ldots)$.

- *Reduce design variable complexity* by displaying the design variables in order from highest to lowest impact (sensitivity) on $margin_{overall}$. Also include high impact variable groups, for cases when interactions among design variables are significant. This allows the designer to focus on the highest-impact variables or variable groups first, and then change lower-impact variables or groups as needed.

We have described how to reduce complexity of the visualized data on three fronts: corners, outputs, and design variables. Now, we bring these three fronts together into an interactive exploration framework, as follows:

- *Relate ordered design variables to overall margin values*: This brings all the pieces together visually. We make each design variable or group in the ordered list *selectable*. Then, if a design variable is selected, plot a 1-D sweep of $margin_{overall}$ versus design variable value. If a design group is selected, plot a 2-D sweep (contour plot) of $margin_{overall}$ versus the two (or more) design variables.

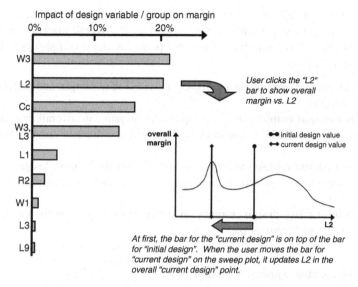

Fig. 6.4 Conceptual example of an interactive design exploration interface. It addresses complexity of several corners, several outputs, and several design variables/groups

- *Allow interactive design changes on sweep plots*: Allow the designer to explore designs directly by changing a "current design" value shown on the 1-D sweep plots and 2-D contour plots.

6.4.2.3 Conceptual User Interface

Building from the interactive exploration framework just described, Fig. 6.4 illustrates a conceptual user interface. It includes the complexity-reducing elements on all three fronts. We now explain the figure.

Figure 6.4 left has a bar plot, where each design variable or variable group corresponds to one horizontal bar. The bars are ordered top-down, starting with the variable or group that has highest impact on the overall margin. In this case, *W3* has the highest impact, *L2* has second-highest-impact, and so on. Note that the group (*W3, L3*) has a high impact. This can occur, for example, when a transistor ratio (*W3/L3*) has high impact.

If the user clicks on the bar for *L2* (Fig. 6.4 center), another plot shows the curve of overall margin versus *L2* (Fig. 6.4 right). Of course, the user can select any bar, to select any design variable or group of design variables, but the impact-based ordering makes it easy for the user to focus on the most important variables.

The plot of overall margin versus *L2* (Fig. 6.4 right) shows specific design points as vertical lines. There is a vertical line for the initial design point, or to be specific, the value of *L2* in the initial design point. There is another vertical line corresponding to the current design. This is the design point that the designer changes in order to improve the overall margin. The user can interactively change the current design value for *L2* on this plot. We see how the user has moved it to a new value away from the initial design, such that the overall margin is higher.

The conceptual user interface of Fig. 6.4 performs visual complexity reduction in a targeted fashion:

- **Reduces corner complexity** by displaying worst-case margin and performance across corners
- **Reduces output complexity** by compressing outputs into overall margin
- **Reduces design variable complexity** by displaying impacts using ordered bar plots
- **Relates ordered design variables to overall margin values**; design variables on the bar plot are selectable, and selecting one plots overall margin versus design value
- **Allows interactive design changes on sweep plots** by moving the sweep plot's "current design" vertical bar

6.4.2.4 Integrated Approach Design Flow

The last section introduced a concept for an interactive design exploration interface (Fig. 6.4). It reduces the complexity of multiple corners, outputs, and design variables and groups by focusing towards design variables and groups that have high impact on the overall (worst-case) margin.

Given this interface, there is a simple flow for users to achieve designs with high overall margin in a small amount of time, yet gaining a large degree of design insight and maintaining full control. The user starts by changing the highest-impacting variable or group to maximize margin. He then proceeds to the next highest-impacting variable or group and changes it to maximize margin. He repeats down the list of design variables or groups, until he is satisfied with the overall margin.

There are many other possible flows. Off the baseline flow, the user may focus on a specific area for more information, such as visualizing the response across each corner, the response of any output, and the effect of any design variable; or even deeper yet such as with viewing waveforms or operating point analysis. The user does not need to follow the streamlined path focusing on the variables impacting overall margin; for example he can focus on variables impacting a single output, while still tracking the effect on other outputs.

6.5 Conclusion

This chapter explored three high-level approaches to front-end design, with emphasis on device sizing in the context of an accurate corner-based design flow.

We first discussed manual approaches, which are familiar to designers, support variation-aware design, and most importantly, allow designers to maintain insight and control over their circuits.

We then considered automated optimization approaches, which can improve sizing turnaround time and help find optimal-quality designs, but compromise insight and control.

Finally, we discussed the concept of an integrated approach, which starts with manual design but adds a highly interactive design exploration interface. It improves sizing turnaround time and helps find optimal-quality designs, yet helps designers to build more insight into the design-performance relation and leaves full control in the designer's hands.

References

Boyd S, Vandenberghe L (2004) Convex optimization. Cambridge University Press, Cambridge

Drennan PG, Kniffin ML (2006) Implications of proximity effects for analog design. In: Proceedings of custom integrated circuits conference (CICC), September 2006

Ferent C, Doboli A (2011) A symbolic technique for automated characterization of the uniqueness and similarity of analog circuit design features. In: Proceedings of design automation and test in Europe (DATE), March 2011, pp 1212–1217

Gielen GGE, Sansen W (1991) Symbolic analysis for automated design of analog integrated circuits. Springer

Graeb HE, Zizala S, Eckmueller J, Antreich K (2001) The sizing rules method for analog integrated circuit design. In: Proceedings of international conference on computer-aided design (ICCAD), pp 343–349

Hansen N, Ostermeier A (2001) Completely derandomized self-adaptation in evolution strategies. Evol Comput 9(2):159–195

Hershenson M, Boyd SP, Lee TH (2001) Optimal design of a CMOS op-amp via geometric programming. IEEE Trans Comput Aided Des Integr Circuits Syst 20(1):1–21

Kirkpatrick S, Gelatt CD, Vecchi MP (1983) Optimization by simulated annealing. Science 220(4598):671–680

Kolda TG, Lewis RM, Torczon V (2003) Optimization by direct search: new perspectives on some classical and modern methods. SIAM Rev 45(3):385–482

Krasnicki M, Phelps R, Rutenbar RA, Carley LR (1999) MAELSTROM: Efficient simulation-based synthesis for custom analog cells. In: Proceedings of design automation conference (DAC), pp 945–950

Magma design automation (2012) Titan ADX Product Page, http://www.magma-da.com/products-solutions/analogmixed/titanADX.aspx. Last accessed 21 May 2012. Magma is now part of Synopsys, Inc

Maricau E, Gielen GGE (2010) Efficient variability-aware NBTI and hot carrier circuit reliability analysis. IEEE Trans Comput Aided Des Integr Circuits Syst 29(12):1884–1893

McConaghy T, Gielen GGE (2009a) Template-free symbolic performance modeling of analog circuits via canonical form functions and genetic programming. IEEE Trans Comput Aided Des 28(8):1162–1175

McConaghy T, Gielen GGE (2009b) Globally reliable variation-aware sizing of analog integrated circuits via response surfaces and structural homotopy. IEEE Trans Comput Aided Des 28(11):1627–1640

McConaghy T, Palmers P, Gao P, Steyaert M, Gielen GGE (2009) Variation-aware analog structural synthesis: a computational intelligence approach. Springer, NY

Mehrotra A (2000) Noise analysis of phase-locked loops. IEEE Trans Circuits Syst I Fundam Theory Appl 49(9):1309–1316

Meissner M, Mitea O, Luy L, Hedrich L (2012) Fast isomorphism testing for a graph-based analog circuit synthesis framework. In: Proceedings of design automation and test in Europe (DATE), March 2012, pp 757–762

Nocedal J, Wright S (1999) Numerical Optimization. Springer-Verlag, NY

Palmers P, McConaghy T, Steyaert M, Gielen GGE (2009) Massively multi-topology sizing of analog integrated circuits. In: Proceedings of design automation and test in Europe (DATE), March 2009

Rutenbar RA (2006) Design automation for analog: the next generation of tool challenges. In: Proceedings of international conference on computer-aided design (ICCAD), pp 458–460

Rutenbar RA, Gielen GGE, Antao B (eds) (2002) Computer aided design of analog integrated circuits and systems. IEEE Press and Wiley-Interscience, NY

Rutenbar RA, Gielen GGE, Roychowdhury J (2007) Hierarchical modeling, optimization and synthesis for system-level analog and RF designs. Proc IEEE 95(3):640–669

Sansen W (2006) Analog Design Essentials. Springer, Dordrecht, The Netherlands

Chapter 7
Conclusion

Variation-aware design can be both fast and accurate. The key is to be able to pair an appropriate variation-aware methodology with the design situation. To do so requires that designers are aware of the options available and have access to a suite of powerful variation-aware tools in their toolboxes. This enables the designer to choose the correct variation-aware approach for the job, whether it is to efficiently achieve comprehensive PVT corner coverage, to design precisely for 3-sigma statistical design, or to achieve true high-sigma statistical designs without unnecessarily spinning silicon or over-margining. Being able to select and apply an appropriate methodology is an important component of being able to deliver competitive, high yield products reliably and on time.

This book presented a suite of variation-aware design methodologies and tools proven to be effective through commercial application. The main topics include:

- *Variation-aware design foundation*: **Chapter 1** reviewed some of the most important concepts that form the basis for all variation-aware design. This includes types of variables and variation, useful variation-aware design terminology, and an overview and comparison of high-level design flows, including the fast, accurate variation-aware design flow promoted by this book.
- *Fast PVT corner extraction and verification*: **Chapter 2** described and compared a suite of approaches and flows for PVT corner-driven design and verification. It then presented Fast PVT, a novel, confidence-driven global optimization technique for PVT corner extraction and verification that is both rapid and reliable.
- *Primer on probabilities*: **Chapter 3** presented a visually-oriented overview of probability density functions, Monte Carlo sampling, and yield estimation. These concepts serve as a foundation for understanding statistical design methods and for interpreting statistical results.
- *3-sigma statistical corner extraction and verification*: **Chapter 4** reviewed a suite of methods used for 2-3 sigma statistical design. It then presented a novel sigma-driven corners flow, which is a fast, accurate, and scalable method

T. McConaghy et al., *Variation-Aware Design of Custom Integrated Circuits: A Hands-on Field Guide*, DOI: 10.1007/978-1-4614-2269-3_7,
© Springer Science+Business Media New York 2013

suitable for 2-3 sigma design and verification. Components of this flow include Optimal Spread Sampling, sigma-driven corner extraction, fast and accurate iterative design over 3σ corners, and confidence-driven 3σ verification.

- *High-sigma statistical corner extraction and verification*: **Chapter 5** reviewed and compared high-sigma design and verification techniques. It then presented a novel technique for high-sigma statistical corner extraction and verification and demonstrated its fast, accurate, scalable, and verifiable qualities across a variety of applied cases. Last, it presented full PDF extraction and system-level analysis.
- *Variation-aware design*: **Chapter 6** compared manual design, automated sizing, and introduced an integrated approach to aid the sizing step in PVT, 3σ statistical and high-sigma statistical design.

Semiconductor profitability hinges on high yield, competitive design performance, and rapid time-to-market. For the designer, this translates to the need to manage diverse variations and to reconcile yield with performance, all while under intense time pressures.

This book is *a field guide* to show how to handle variation proactively, and to understand the benefits of doing so. Ultimately, it is a distillation of successful and not-so-successful ideas, of lessons learned, all geared towards delivering *better designs despite variation issues*.

References

The ideas in this book are implemented in a commercially available tool (Solido Design Automation, 2012). Several of the core ideas have patents and patents pending, as follows.

McConaghy T et al (2007) System and method for determining and visualizing tradeoffs between yield and performance in electrical circuit designs, USPTO patent #7,689,952

McConaghy T et al (2007) Data-mining-based knowledge extraction and visualization of analog/mixed-signal/custom digital circuit design flow, USPTO patent #7,707,533

McConaghy T et al (2007) Interactive schematic for use in analog, mixed-signal, and custom digital circuit design, USPTO patent #7,761,834

McConaghy T et al (2007) Model-building optimization, USPTO patent #8,006,220

McConaghy T et al (2008) Global statistical optimization, characterization, and design, USPTO patent #8,024,682

McConaghy T et al (2009) Pruning-based variation-aware design, USPTO patent #8,074,189

McConaghy T (2007) Modeling of systems using canonical form functions and symbolic regression, USPTO filing #20070208548

McConaghy T et al (2008) On-the-fly improvement of certainty of statistical estimates in statistical design, with corresponding visual feedback, USPTO filing #20080300847

McConaghy T et al (2009) Proximity-aware circuit design method, USPTO filing #20110055782

McConaghy T (2009) Trustworthy structural synthesis and expert knowledge extraction with application to analog circuit design, USPTO filing #20090307638

McConaghy T (2010) Method and system for identifying rare-event failure rates, USPTO filing #61/407,230

McConaghy T (2011) Monte-Carlo based accurate corner extraction, USPTO Provisional Filing

McConaghy T (2011) Fast function extraction, USPTO Provisional Filing

McConaghy T (2011) Method for fast worst-case analysis, USPTO Provisional Filing

Solido Design Automation (2012) Variation Designer toolset. http://www.solidodesign.com